AMERICAN ISSUES

The Environment

AMERICAN ISSUES

The Environment

Gerald Leinwand

Facts On File
New York • Oxford • Sydney

The Environment (American Issues Series)

Facts On File, Inc.
460 Park Avenue South
New York NY 10016
USA

Facts On File Limited
Collins Street
Oxford OX4 1XJ
United Kingdom

Facts On File Pty Ltd
Talavera & Khartoum Rds
North Ryde NSW 2113
Australia

Produced by the Philip Lief Group

Library of Congress Cataloging-in-Publication Data
Leinwand, Gerald.
The environment / Gerald Leinwand.
p cm. — (American issues)
Includes bibliography references.
Summary: Surveys environmental problems in the United States, examines legal and social aspects of environmentalism, and discusses political plans in place to deal with the environmental decline.
ISBN 0-8160-2099-X
1. Pollution—United States—Juvenile literature. 2. Environmental protection—United States—Juvenile literature. 3. Environmental policy—United States—Juvenile literature. [1. Pollution. 2. Environmental protection. 3. Environmental policy.] I. Title. II. Series: American issues (New York, N.Y.) TD180.L45 1990
363.7'00973—dc20 89-77351

A British CIP catalogue record for this book is available from the British Library.

Australian CIP data available on request from Facts On File.

Jacket design by Catherine Hyman
Composition by Facts On File, Inc.
Manufactured by Maple-Vail Manufacturing Group
Printed in the United States of America

10 9 8 7 6 5 4 3 2 1

This book is printed on acid-free paper.

CONTENTS

PREFACE

How shall we care for the environment? Your future is closely linked to this question—and to the answers.

Although science and technology appear to provide us with unlimited opportunities to change the environment, the extent of human control is limited.

What are those limits? How will we recognize them? What are the dangers if we try to go beyond those limits, or carelessly pollute our landscape? How can resources be managed better so that the future needs of men and women will be met?

Once you start thinking about these questions and the possible answers to them, you are assuming the responsibilities involved in the stewardship of planet Earth. That stewardship has two parts. It involves tending your own garden, that is, caring for the American landscape; it also encompasses caring for the global landscape, humanity's collective home.

The Environment is a book designed to help you become a participant in the search for answers to some of these questions. The presentation of environmental issues here should help you carry out your stewardship responsibilities more effectively.

The book is one of a series called *American Issues*. This series probes contemporary issues, examines possible solutions and helps you begin to think about new directions and alternatives to problems that do not lend themselves to easy answers.

Gerald Leinwand
New York, New York

CHAPTER ONE

The American Landscape: *A Historical Overview of Ideas and Attitudes*

"Land Ho!" cried the bleary-eyed sailor who, through a telescope held up to one eye, sighted the mist-shrouded shores of the New World. Beyond the headlands and the beaches, he must have wondered, what kind of land was it? The courage of those 17th-century explorers and of those who sailed with them cannot be overestimated. They came to establish colonies, the first of which would be called Jamestown (in what is now Virginia), built in 1607, and Plymouth (in what is now Massachusetts), founded in 1620. The seafarers would need every ounce of their courage to tame the land in the New World, for it was not only unfamiliar in its flora (plants) and fauna (animals), but vast beyond imagining.

Not knowing the land, the early settlers at first feared it as much as they had the sea. The land's interior was remote, wild and mysterious. When the early colonists began to feel at home in the New World, they became aware of its great beauty and spectacular abundance. As the American landscape revealed itself to them, they were in turn challenged by its extent and variety, and thrilled with its beauty and bounty. Ultimately, they would adapt to it, and the process would yield the unique American character. But in doing so, they would despoil the very land upon which they depended.

The Land

In the New World, the early settlers found a primeval forest that dominated an area as extensive as the ancient Roman Empire. Europe had not known a forest so vast for 2,000 years. From Maine to Florida, the colonists were confronted with dense woods nearly everywhere. So thick were the branches of trees and so luxurious the foliage, a squirrel could leap from branch to branch for a thousand miles without touching the ground or seeing the sunlight.

In the beginning, the forests were an obstacle to be overcome. The settlers had to clear them in order to claim the fertile soil that lay beneath. But the forests were also valuable, for they supplied logs for building houses as well as firewood for heating those houses; wood for making furniture and lumber for wagons; and timber, pitch and tar for building ships. The forest provided berries, nuts, maple sugar, honey and wild game for food provisions. The pelts of beavers, raccoons, deer and bear that lived there offered warmth and were a valuable resource for trade.

As the land became more familiar, its grand design was easier to see. The narrow coastal plan gave way to the rolling peaks of the Appalachian Mountains. Between that mountain range and the mighty Mississippi River lay an area described by historian James Truslow Adams as "the most spacious habitation for human life to be found in the world."[1]

Beyond, from the Mississippi Valley in the East to the Rocky Mountains in the West, stretched broad grasslands known as the Great Plains. Lush grazing lands, the Great Plains were the home of huge herds of buffalo before European settlers arrived and hunted them almost to extinction. The Rockies rose above the western edge of the Great Plains, forming the eastern rim of the Great Basin, whose western edge was formed by the Sierra Nevada and the other coastal mountain ranges. A strip of fertile flatlands made up the continent's western coastline, which was washed by the waters of the Pacific Ocean.

It would take the newcomers some time to discover, but deep in the American earth lay vast mineral wealth. There were silver and gold, to be sure, but also coal and oil, which would prove to be valuable sources of energy. The continent's soil also contained the rare minerals that would one day serve as fuel for nuclear power plants. While it did not have everything, the New World had ample resources for a new and growing nation.

Water

Newcomers to North America found deep harbors along the Atlantic seaboard, where later such great port cities as Charleston (South Carolina), Baltimore (Maryland), New York (New York) and Boston (Massachusetts) were founded. In the coastal waters, great schools of fish—sturgeon, catfish, bass, salmon, mackerel—thrived. "The abundance of Sea-Fish are almost beyond beleeuing," according to a New England settler quoted in Daniel Boorstin's *The Americans: The National Experience.* "And sure I would scarce have beleeued it except I had seene it with mine owne Eyes."[2] Codfish was especially plentiful in New England, and gave its name to Cape Cod, the Massachusetts peninsula.

Between Canada and the United States was the finest route to the Great Plains, the St. Lawrence River. It connected the Great Lakes to the Hudson and Mohawk rivers and so permitted passage into the heart of the continent. The Great Lakes, covering nearly a hundred thousand square miles, held half of the world's supply of fresh water. The great river system of the Mississippi, Missouri and Ohio rivers allowed travel between the Gulf of Mexico and the Great Lakes, a distance of 2,000 miles. The Colorado River tumbled down the Rocky Mountains and flowed through the Grand Canyon before emptying into Mexico's Gulf of California. To the northwest, the Columbia River and its important tributaries drained into the Pacific from the Cascade Mountains. The river system of America offered a relatively easy avenue of exploration, travel, communication and trade. It also watered the continent's fertile soils and helped make them abundantly productive.

Air

The varied American landscape was matched by an equally variable climate. The early settlers found the climate in the New World both colder and hotter than that to which they were accustomed. The settlers in Plymouth soon discovered that the Massachusetts winters were colder (and that rain was more limited) than in England. Those who settled south of the Chesapeake Bay found that the climate was warmer. In nearly all parts of the new land, however, the summers would be much hotter than those of Europe. In warmer climes, killing frosts were possible, while debilitating heat sometimes seared normally cooler regions.

Nowhere on the new continent did the climate inhibit industrial and agricultural development. There was no cold and barren region like Siberia in Russia, nor were there the impenetrable tropical jungles that hampered development such as those in Brazil. As the settlers learned more about the climate, they found that North America was indeed a world of contrasts.[3] They discovered the tropical, moist air of the Gulf coasts and the dry air of the high western plateaus. They encountered the cruel winters of Maine and the blizzards that swept the Great Plains. From the rolling hills of New England to the flat plains of the Midwest; from the jagged peaks of the western mountain ranges to the arid deserts of the Southwest, the land was almost limitless in its variety.

The early settlers did not realize how far the Pacific Ocean lay from the Atlantic. Indeed, they expected a passage to it would soon be found, a presumption that they used as a promise to beckon other settlers. Little did the early colonists know that their enterprise—founding a European outpost in an uncharted land—would become one of the world's largest nations with a greater global influence than they could then imagine. But as the secrets of their new home revealed themselves, the settlers realized the potential of the land they had come to make their own.

The American Landscape and the American Character

If history is drama, then the environment is the stage. One of the central themes in American history is the story of how men and women have faced the challenges of living in a wilderness and taken advantage of nature's abundance for food, clothing and shelter. In the course of this struggle, the American character was forged.

Theories about the relationship of the American landscape to the American character and form of government abound. In 1796, an English visitor observed that the New World's abundance helped create the independent spirit of the American settler. His theory held that, because each person could easily live off the fat of the land, no one could be controlled by those who possessed more than he or she did.

Economist Henry George believed that nearly every aspect of the American character, including the American genius for invention, level of intelligence, power of adaptation, and independent and optimistic spirit, had its basis in the availability of vast amounts of open land.[4] Free or inexpensive land also played a central role in fostering loyalty to the new nation among its citizens. When a small population inhabits a large expanse of desirable land, each person has access to more than enough land to support him or herself. No one need be poor, and no one can gain too much power over others. Abundant land thus helps a democratic form of government develop.

President Woodrow Wilson attributed the greatness he saw in the American democracy to the size and character of the continent that Americans made their own. Sharing Wilson's view, historian Max Lerner wrote, "The wild and lush landscape seemed to offer to the eighteenth-century mind, which personified Nature with a capital N, a charter to build on the far-stretched continent a new social system swept clean of privilege, caste, and inequity."[5]

Americans could not help but be shaped by the bounty of the land, and by the assumption that here was a land of unlimited natural wealth. Generations of white settlers and their descendants proceeded as if the resources of North America were essentially inexhaustible. Not until the late 19th century did they realize that nature needed to be protected from humanity.[6]

From earliest days Americans have wasted their resources. To the first European pioneers, the forests were an obstacle to settlement and farming. The settlers cut down huge trees, burned the forests, plowed the soil, and moved on to virgin lands elsewhere. One-crop farming, in which land was devoted exclusively to cotton, wheat or corn, ate up the nutrients of the soil. Overgrazing of cattle and sheep on the Great Plains left those lands vulnerable to erosion. Without the benefit of trees and grasses, once-fertile earth thus denuded was literally blown away during recurrent dust storms, or washed away by rains.

tlers sought to buy the land: "Sell the country?...Why not sell the air, the clouds, the great sea?"[8] The New World, as early colonists found it, was a tribute to the care with which the Indian treated the environment.

Yet, American Indians were not ideal environmentalists. Their slash-and-burn techniques of clearing the forests for farming were wasteful and destructive. Overhunting and primitive methods of farming clashed with their generally more reverent approach to their environment.

Similarly, some white pioneers were aware that the bounty they found in their new home might not be inexhaustible. They soon recognized their dependence upon the forests for so many of their needs. The colonists understood that the forests were precious, supplying them with a great deal of lumber for heating, cooking, building and other purposes. Timber was also an important item in overseas trade, and the glass- and metal-making industries depended on wood to fuel their fires. The importance of forests to the early American settler can also be seen in the abundance of wooden household items: rakes, shovels, pails, ladles, bowls, spoons, washboards, tubs and even nails.

The settlers came from European countries that had already experienced shortages of timber, their forests and soil having been made unproductive from overuse. Some of these countries had regulations governing the use of land and trees, which the early colonists adapted to their environment. Just six years after the founding of Plymouth Colony in Massachusetts, the community ruled that no timber could be cut on colony lands without official consent. Within 50 years, nearly every colony in the New World had similar restrictions. In 1681, Pennsylvania ruled that one acre of trees be left standing for every five cleared. The British government also imposed restrictions on what could be done with colonial forests. It was especially interested in white pine, a type of lumber already in short supply in Europe. The very tall, straight trees made excellent masts for ships, and were absolutely essential for Britain's efforts to establish naval superiority.

In 1710, as a forerunner of what would later become a vast system of state and national parks, the community of Newington, New Hampshire, set aside 110 acres as a community forest. A similar community forest was established in Danville, New Hampshire, in 1760. When the colonies became states of the new nation, they took the initiative to try to preserve these forests, as well as to protect fish and wildlife and curb water pollution.

The role of the federal government in conservation was at first limited, but began with concern for the nation's forests. In 1799, through the Federal Timber Purchases Act, Congress appropriated $200,000 to buy timber for American ships. As the American navy expanded after the War of 1812, the government confirmed its interest in the nation's forests as essential to its need for naval and merchant vessels. Among the early presidents, John Quincy Adams was particularly concerned with the relationship of the environment to continued national development. In 1827, he took steps to define a policy on public land and its resources by setting aside 30,000 acres of live oak in the Bay of Pensacola in Florida. The Santa Rosa Reservation, as it was called, was also the site of experiments, using varied soil, water and cultivation techniques, to improve the quality of the oak used in battleships. The Santa Rosa Reservation did not last much beyond the Adams administration, yet it established a pattern for public land use that would later be followed.

In 1849, the Department of the Interior was established, with responsibility for all United States national parks. Then, in 1872 the federal government created Yellowstone National Park on 3,200 square miles of pristine Wyoming land. A year later, Congress enacted a plan for distributing land to western settlers. Each homesteader would receive 160 acres of land in the Great Plains by agreeing to plant one quarter of it with trees.

In 1872, April 10 was declared Arbor Day, a special time for the planting of trees. On September 10, 1875, the American Forestry Association was organized to make Americans more sensitive to the importance of timber cultivation and to lobby Congress on behalf of the forest industry. Carl Schurz, a German immigrant, became secretary of the interior and proposed the establishment of federal forest reservations and the use of scientific techniques in the care of forests.

In 1891, Congress gave the president power to establish forest reserves on public land. During the course of his administration, President William Henry Harrison set aside forest reservations totaling 13 million acres. President Grover Cleveland added 25 million acres, and President William McKinley, seven million.

The Environment Finds a "Bully Pulpit"

Although both George Washington and Thomas Jefferson warned the nation of the dangers of wastefully destroying its forests or needlessly depleting its soil, American resources were so bountiful that the nearly universal view in the early days of the republic was that they could never be exhausted. Jefferson referred to America as a "Chosen Country with room enough for all descendants to the hundredth and thousandth generation." As a result, citizens were slow to understand the need to protect the environment.

An early pioneer of the conservation movement was George P. Marsh, whose 1864 book, *Man and Nature: or Physical Geography as Modified by Human Action*, was among the first to point out the perils that lay in wanton interference with nature. Out of his concern came a growing public interest in ecology, or the relationship between the various parts of the environment and the human capacity to modify it.

Soon afterward, Major J. W. Powell, director of the U.S. Geological Survey from 1881 and 1894, called public attention to the importance of caring for western lands and to the role proper irrigation could play in making those lands more productive. As head of the United States Forest Service, Gifford Pinchot urged that a portion of the nation's forests be set aside for future use. Pinchot became Theodore Roosevelt's conscience on matters of the environment.

In Theodore Roosevelt, who served as president of the United States from 1901 through 1909, the environment found a passionate spokesman. Roosevelt's interest in the environment went back to his youth. As a young man, he had spent a considerable amount of time on western ranches in a successful effort to overcome the handicap of a frail body. Throughout his life he enjoyed nature and was fond of hunting. Roosevelt used the presidency, which he called the "bully pulpit," to alert legislators to the need to safeguard the nation's environment and to enact environmental policy. Reclaiming arid land by irrigation, conserving water power and other natural resources, and protecting land and timber from exploitation by industry were part of Roosevelt's campaign against abusive corporate giants.

Alone among American presidents up to his time, Roosevelt grasped the relationship of environmental matters to national well-being. In 1907, he appointed an Inland Waterways Commission and assigned it the task of reviewing the relation of rivers, soil, forest, water-power development, and water transportation. On May 13, 1908, he called a governors' conference on the conservation of natural resources. To this conference he invited the vice-president, members of the United States Supreme Court, members of the Inland Waterways Commission, and members of Congress, as well as representatives from environmentally concerned organizations.

As never before, the conference turned the national spotlight on the need for conservation. In its declaration of principles, the conference urged conservation of forests, waters and minerals, and expressed concern about soil erosion and irrigation. It recommended that the government retain control of land containing coal, oil, phosphates, natural gas and other fuel deposits. The conference also noted the need for the improvement of navigable streams and the regulation of timber-cutting on private lands.

As a follow-up to the conference, a number of states established their own conservation commissions. In 1909 the National Conservation Association was organized as a means of educating the American people about the nation's environmental needs. A national commission was established under the leadership of Gifford Pinchot, and was assigned the task of taking an inventory of the nation's natural resources. In recognition that environmental matters were international in character, President Roosevelt sought and obtained the cooperation of other American nations in establishing a North American Conservation Commission.

American naturalist John Muir took a somewhat different approach to the environment from that of Gifford Pinchot. Pinchot was primarily a conservationist, where Muir was mainly a preservationist. While these two approaches to the environment overlap, they sometimes conflict. Conservationists seek mainly to manage scarce resources so that they may be used gradually, over a long period of time, to serve the needs of the greatest number of people. They believe in saving resources today so that they may be used at a later date. In short, the emphasis of conservationists is on future use.

Preservationists, such as John Muir, take the view that huge sections of pristine land should be kept permanently in a natural state. Any tinkering with the environment destroys the harmony of nature. It is not enough to conserve forests for future use—forests and all the wildlife in them must be preserved without any thought of ever using them at all. According to this view, all creatures are fellow travelers on Earth and must learn to share the planet with one another.

The division between preservationists and conservationists endures to this day. The preservationist camp appears to be losing out as it becomes ever more difficult to preserve national parks unscarred by the onrush of humanity. No matter how vast the park, waves of economic development crash against its boundaries, threatening to foul its air with smog and pollute its waters with waste. The drive for economic development and the desire to sustain ever-improving living standards cannot be suppressed, for men and women are entitled to pursue their self-interest in any lawful way. Squaring preservation with development is thus a sort of environmental balancing act, and one that will last throughout our lives.

A New Deal for the Environment Under FDR

In spite of Theodore Roosevelt's substantial environmental accomplishments, a great deal remained undone. Most Americans were incapable of thinking the unthinkable—that, within a surprisingly short period of time, important natural resources would be depleted and the environment permanently damaged unless prudent measures were taken. Inefficient mining wasted millions of tons of coal. Reckless drilling wasted many more barrels of oil and billions of cubic feet of natural gas. Millions of acres of forest were destroyed by forest fires, most of which were set by indifferent or careless campers. Despite the efforts of Gifford Pinchot, protecting the forests from predatory lumber barons remained a struggle. They continued their wasteful harvesting of trees that had taken hundreds of years to mature. Only in our own time, and even now not with total consistency, have lumber developers been more careful in how they harvest the forest and in how they replant to replace what's been cut.

Between the administrations of Theodore Roosevelt and his distant cousin Franklin D. Roosevelt, environmental legislation moved forward at a snail's pace. But under FDR, president from 1933 to 1945, environmental issues became central once again. During his years in office, America sought to work its way out of the Great Depression, which lasted roughly from 1929 to 1941. Roosevelt liked to think of himself as a farmer and, indeed, his knowledge of the land was notable. "The forests," he said, "are the 'lungs' of our land, purifying our air and giving fresh strength to our people."

It came almost as second nature for FDR to view conservation as an element in his attack on the Great Depression. As part of his "Hundred Days" program, named for the extremely productive first three months of his administration, he established the Civilian Conservation Corps, designed to put the unemployed to work by conserving America's forests and soils. By the time the CCC, as it was popularly called, was abandoned during World War II, 2.5 million men had participated in it. Its accomplishments were impressive. Seventeen million acres of new forest had been planted, state and national parks had been groomed, dams and fish ponds had been built, Dutch elm disease had been brought under control, historic battlefields had been restored, beaches had been cleared, and campgrounds and recreational sites had been improved.

Under Roosevelt, tree planting became a high priority through the country as one way to help eradicate the "dust bowl" that drought and poor agricultural practices created in the 1930s. Roosevelt urged the National Forest Service to plant a "shelter belt" of trees creating a natural barrier one hundred miles wide and stretching along the 100th meridian from Canada to Abilene, Texas. This would help bind the soil by breaking the wind and catching the rain and snow. With the help from the CCC, the Forest Service planted 200 million trees to restore fertility to the dust bowl. In 1935, Congress passed the Soil

Conservationist Act, making the conservation of soil a national responsibility.[9]

The crowning achievement of Roosevelt's conservation efforts was undoubtedly the Tennessee Valley Authority. As far back as 1825, John C. Calhoun launched a survey of the area around the town of Muscle Shoals, Alabama, to determine the merits of building a canal around the rapids (called shoals) on the Tennessee River. Although early attempts to build the canal were unsuccessful, by 1890 the United States Corps of Engineers had successfully completed it. In 1917, President Wilson chose Muscle Shoals as the site for a government-owned dam that would generate hydroelectric power for a munitions plant there.

After World War I, Senator George Norris of Nebraska persuaded Congress not to sell the facilities at Muscle Shoals. Senator Norris hoped to establish a regional approach to the energy needs of the 40,000-square-mile area drained by the Tennessee River.

The people of the Tennessee Valley, which included parts of North Carolina, Virginia, Tennessee, Alabama, Mississippi, Kentucky and Georgia, were appallingly poor, beset by hunger, and riddled with disease. They were to be the beneficiaries of Norris's project. A system of dams would hold back the unruly waters of the Tennessee (which flooded annually) and provide these residents with the electricity they lacked. The plan would also improve navigation, help reforest the region, produce fertilizer, offer recreational opportunities and encourage the general economic development of the area. The project was finally launched during Roosevelt's administration, as part of the National Recovery Act.

The career of the TVA was marked by bitter controversy. There were those who condemned it as socialism, for it kept private interests from providing electricity. Others felt that the TVA was guilty of racial discrimination in its personnel practices. In later years, it became the target of conservationists who thought its use of coal in producing electricity was contributing to environmental pollution. Nonetheless, the TVA was a model of a sensible regional approach to environmental concerns. Its future may be unclear, but its past was marked with success.

Through science and technology we have made more of an impact on our environment in the past 100 years than was made in the preceding 3,000. But are there limits to what the Earth can sustain? Can we continue to improve crop yield by adding unlimited amounts of chemical fertilizer to our soil? If you and your children are to inherit the Earth, what kind of Earth will it be? These are some of the questions we will examine in the next chapter.

Notes

Citations in the notes are brief. Full citations appear in the bibliography.

CHAPTER ONE

1. Adams, 4.
2. Boorstin, 6.
3. Adams, 5–6.
4. George.
5. Lerner, 104.
6. Lerner, 107.
7. White, 25.
8. Udall, 31.
9. Schlesinger, 68.

CHAPTER TWO

The Good Earth: *Land Pollution*

Wednesday, April 22, 1970, was the day the Earth stood still. It stood still not because of any cosmic phenomenon but because men and women throughout the United States took time out from their daily activities to express their concern about what was happening to the Earth on which they lived. It was called Earth Day.

Congress stood in recess because so many of its members were taking part. Ten thousand people joined in a rally at the Washington Monument in the nation's capital; huge rallies of up to 25,000 people were held in New York, Philadelphia, Chicago and other big cities. Two thousand colleges commemorated the day, and 10 million schoolchildren in 10,000 elementary and high schools participated in "teach-in" programs on the subject of the environment. John Lindsay, then mayor of New York City, set the tone for the day. In a speech quoted in the *New York Times* the following day, he said, "Behind the complex predictions and obscure language, beyond words like ecology, environment and pollution there is the simple question: Do we want to live or die?"[1] Twenty years later, that is still the question, and one that was addressed on April 22, 1990—Earth Day 1990. A strong answer came twenty years to the day of the first Earth Day celebration. On April 23, 1990, across America and all over the world, people gathered for another Earth Day celebration to show that they cared. This time, over 750,000 gathered in New York's Central Park. But support was widespread. Earth Day 1990 organizers said that over 200 million people in 140 nations had taken part in the largest grassroots demonstration in history.

It was a day people spent picking up trash on a beach, planting a tree, attending workshops and gatherings to learn how they could get involved to ensure the future of the Earth. People learned that even small things that they can do can go a long way to a better environment.

There Is No Free Lunch

Since World War II, under the impact of new technology, growing populations, higher living standards and rapid utilization of natural resources, the life-sustaining balance between humanity and nature has been breaking down. The science of how that balance can be maintained and how it is threatened is called ecology. More formally, ecology is the study of the relationship between organisms, including people, and their environment. Barry Commoner, America's best-known ecologist, defines ecology as the "science of planetary housekeeping."[2]

Commoner identified four laws of ecology:

1. *Everything is connected to everything else.* A cow, for instance, depends on plants for food. People in turn depend on cows for milk and meat. The cow's manure enriches the soil and improves the quality of the grass. The richer the grass, the higher the quality of cow's milk and meat. However, if the grass

the cow eats becomes contaminated, say by chemical fertilizers, the quality of the food we get from cows is endangered and our health is threatened.

2. *Everything must go somewhere.* When it breathes, the cow gives off carbon dioxide, an important nutrient for plants. Plants give off oxygen, which animals breathe. Other waste products, however, are not so useful. Where, for example, does the used battery in your flashlight go when you throw it out? You probably toss it into a garbage can. The trash in the can is then picked up by a sanitation truck, and perhaps taken to an incinerator. When the battery is burned, a toxic (poisonous) vapor is produced. It is released into the air and carried off by the wind. Eventually it returns to the Earth dissolved in rain or snow. If it falls into a mountain lake, it may be eaten by fish which you, in turn, might catch and eat. The toxic waste from your battery, now in your system, may be harmful to your health.

3. *Nature knows best.* This law contradicts what modern men and women have been led to believe, namely, that science and technology can improve upon nature. Often they can, but not without some negative effects. Every artificial substance introduced into the environment is likely to cause some ecological imbalance. On September 21, 1988, for example, the *New York Times* published an article on the safety of eating grapefruit treated with chemicals. The article pointed out that the federal government allows grapefruit growers to spray their trees with lead arsenate (to hasten ripening) even though scientists had warned of its dangers. The National Coalition Against the Misuse of Pesticides had publicized an internal memo of the Environmental Protection Agency stating lead arsenate is a carcinogen. The articles show how tinkering with nature may be hazardous to your health.

4. *There is no such thing as a free lunch.* This last ecological law summarizes the previous three. Because everything in nature

is connected with everything else, every gain is won at some cost.[3]

Think of yourself as a traveler on the spaceship Earth. Everything on board must work in harmony if the spacecraft is to be kept on course. This is the meaning of ecology. To the extent that each person is responsible for making sure that the spacecraft is not annihilated, each must care for our home in space. The care of its soil, water and air becomes your responsibility.

And No Birds Sing

In 1962, Rachel Carson published *Silent Spring*. It became one of the most influential books in the literature of environmental concerns. In it Carson warned that the excessive use of pesticides was poisoning the food we eat, the flowers we grow, the air we breathe. The result, she said, is the destruction of the world where we live.

Her book begins with a "A Fable for Tomorrow," the story of an idyllic American town whose entire human and animal population was suddenly overtaken by sickness and death. The fable concludes with the somber observation that the people of the town had brought this tragedy upon themselves. A fable is a short story that teaches a moral lesson. What can we learn from Carson's fable?

What we learn is that farmers, seeking to improve their crops, have ignored the consequences of using chemicals. In using pesticides (the term we will use for all of the chemical additives used in gardens, on farms and in animal feed), they hoped to ripen their food more quickly, increase their harvest, grow bigger and fatter livestock, and make food more plentiful at ever-decreasing costs while enchancing their own profits. For all their dangers, pesticides have contributed mightily to the nation's prosperity, to the productivity of farmers, to the bounty, variety and modest cost of our food, and to our ability to share our food surplus with less fortunate people.

But in this process there is a high price to pay. The term "pesticide" itself provides some clues about the dangers of using them. The word comes from the Latin *pestis*, which means "plague," and *cida*, which means "to kill." Pesticides are chemical agents meant to kill insects, plants, fungi, mites, rodents, bacteria and a great many other pests that endanger crops. Since the publication of *Silent Spring*, the sale of pesticides has more than doubled. Registered with the Environmental Protection Agency are 34,000 pesticides made from 600 basic chemicals. These pesticides need to be handled and used with extreme care. While they kill pests, they can kill people as well. They can improve farm productivity, but they can contribute to the destruction of the environment. Used in excess and without adequate care, they can destroy the earth's ecology. On November 14, 1988, the *New York Times* summarized the dangers from pesticides as follows:

1. Pesticides contaminate groundwater, rivers and estuaries throughout the country.
2. Pesticides can kill or sicken farm workers who apply them. In a National Cancer Institute study, farmers in Kansas were found six times more likely to contract lymphatic cancer if exposed to herbicides more than 20 days a year.
3. Pesticide residues often contaminate the crops they are meant to protect. Americans are exposed to pesticide residues in almost everything they eat—meat, dairy products, fruits, vegetables, sugar, coffee, oils and most processed foods.
4. Pesticides too often provide only temporary relief. They must be increased or replaced as pests develop resistance.[4]

Thus, to grow enough high-quality food, it is unlikely that farmers or ranchers can do without chemicals altogether. But if pesticides are to be used in the future, as is very likely, there are many questions that need to be answered. What kind of pesticides are the safest? In what quantity should they be used? Should some be banned altogether? If so, which ones? For those that cannot be banned, how can they best be controlled? Are there alternatives to the use of pesticides?

In 1971, one of the earliest acts of the Environmental Protection Agency was to ban DDT, a pesticide that Carson called an "elixir of death." DDT is a chlorinated hydrocarbon compound, which, when first introduced, resulted in the saving of millions from starvation, malaria and typhus. But DDT can kill a broad range of living organisms. In addition to killing some of the pests it was intended to kill, it killed wildlife such as bald eagles, peregrine falcons and brown pelicans. DDT residues lingered in the environment, too, and, in some cases, killed fish and caused cancer in people.

The Environmental Protection Agency is charged with the responsibility of testing the impact on human health and the environment of the chemicals found in pesticides. The Federal Insecticide, Fungicide and Rodenticide Act of 1971 required the EPA to retest those pesticides that were on the market before 1971, as well as to test all new ones. However, this program of testing proceeds at a snail's pace, and as recently as 1988 only a few products had been tested. The EPA estimates that it will take 20 more years to complete the testing.[5]

The law requires the EPA to make sure that pesticides have no "unreasonable adverse effects" on the health of people or on the environment. Determining what is reasonable and what is unreasonable, however, has thus far proven to be an almost insoluble problem. No pesticide is totally free of potential hazard to the environment or to human health. But in most cases it would be impossible to bring new crops to market at reasonable prices without their use. The general standard adopted by the EPA for a "reasonable" risk of cancer from a pesticide was no more than one case of cancer in every million people exposed over a 70-year lifetime. But is such a standard good enough? Is such a risk worth taking? What would you decide in the following real situations?

The pesticide Dinoseb is used on peas, beans and other crops. In 1986, the EPA decided that Dinoseb posed an unreasonable risk of cancer to farm workers and children, and caused birth defects as well. There was, said the EPA, no "adequate margin of safety" at any level of exposure. Growers who used Dinoseb took the agency to court, claiming that there were no alternative pesticides that they could use, and that withdrawing it from the market would result in severe economic loss. A federal court ordered the agency to modify its decision and allow some limited use of the chemical. Was the court right in its decision?

Let's try another one. Alachlor is the nation's most widely used herbicide (weed killer). The EPA determined that the cancer risk from Alachlor was one in a million over a lifetime, that the chemical was contaminating underground water supplies, and that there were alternative herbicides. But the EPA decided to allow continued use, because the risks to people and the environment were "not unreasonable" in view of the economic benefits for the nation's corn, soybean and other crops.[6] What decision would you have reached?

From these two examples, you can see that making a determination of whether or not to allow farmers, ranchers and gardeners to use a particular pesticide is a difficult process. Ideally, *no* cancer risk would be the yardstick applied. Yet, if that standard were to be adopted, much of what we want to eat would be either inedible, unavailable or too expensive.

In October 1988 the Environmental Protection Agency was given new powers and new responsibilities. Over the next nine years it must test all pesticides. Moreover, the EPA is no longer required to compensate manufacturers for unused inventories of banned pesticides. The costly requirement has made the EPA reluctant to ban chemicals.

This change in the law should mean that by the year 2000, pesticides will be more carefully controlled. Yet the nation has clearly relied upon pesticides heavily, and alternatives still must be sought. Among the possible alternatives is the development of biological approaches. These would produce better crop yields by encouraging the natural predators of destructive insects. If the technique proves successful, some pesticides could be gradually phased out.

Another alternative could be genetic engineering. Pest-resistant strains of plants that require no protection from poisonous chemicals may be developed. But if science unleashes the genetic genie, what other perils may follow? Until we know, pesticides will have to be used. However, since the environmental impact of these chemicals is harsh, and the risks in some cases unacceptable, stringent controls will have to be adopted. The pesticide industry has to be encouraged to regulate itself more carefully.

Soil Erosion

"Every moving thing lifted the dust into the air: a walking man lifted a thin layer as high as his waist, and a wagon lifted the dust as high as the fence tops, and an automobile boiled a cloud behind it." This is how, in his searing novel *The Grapes of Wrath*, John Steinbeck described the plight of a rural Oklahoma family who watched their farm literally blow away in the wind. During the Great Depression of the 1930s, a drought ravaged the Great Plains of North America. Ignorant of how to save their fragile soil, farmers watched as erosion turned their thin layer of topsoil into dust. The wind that swept across the plains stirred up dust clouds so thick sometimes it seemed as if the night would last forever. Dank, dark and heavy dust invaded the ears and nostrils of men, women and beasts. It clung between the toes, filtered under door sills and through cracks in windows, sifted down chimneys, onto the tables, and into stewing pots and soup bowls.

Erosion is the wearing away of the soil by water, wind, waves and glaciers. Water is responsible for two thirds of the erosion of farmland, usually from rain, oceans and rivers. Soil erosion increases the cost of making a living to farmers and ranchers, decreases their profits, depreciates the value of their farmland or rangeland, and increases prices for the consumer. Obviously, without water there could be no agriculture, but farmers and ranchers must learn how to manage their land so as to enjoy the benefits of the water supply while minimizing the loss of topsoil. Billions of dollars are lost each year as a result of ignorance.

Planting the same crop year after year is another major cause of soil erosion. Because the same minerals are depleted from the soil each year, the soil eventually becomes unable to sustain crops. The use of manure or other fertilizers will often diminish erosion, but only for a time. Crop rotation, that is, planting different crops each year, helps the valuable topsoil remain in place. But economic pressures make it difficult for farmers to switch from one crop to another. If there is high demand and thus high profit for one crop but not another, the farmer is likely to grow the same crop year after year.

As the growing populations demand greater supplies of food, there is pressure to put less suitable land into use. But every time such land is put under the plow, the farmer must farm the land more skillfully. Where farming is done on sloping land, for instance, rapid erosion may occur unless the farmer knows how to plow in such a way as to catch the needed rainfall. Otherwise, rapid runoff may turn the farmland into gullies that wash away the topsoil. Cultivating sloped land involves choosing the right crop, applying the appropriate fertilizers, and using sophisticated forms of plowing. For example, the farmer can plow in steps (terracing) or in curved patterns (contour plowing). In another technique, fallowing, the land is allowed to lie barren for a season in order to accumulate moisture. However, in order to prevent erosion of the soil left uncultivated during fallowing, farmers must cultivate adjacent strips.

Despite these and other measures, including FDR's program of planting tracts of trees to bind the soil and break the wind, nearly 44% of U.S. cropland still loses topsoil every year. Imprudent farming designed to profit quickly from prevailing market conditions is a major cause of erosion. When important nutrients are recklessly taken from the soil, a renewable resource (soil) is no longer renewable. This form of farming is more like mining—and, indeed, the expression "mining the soil" is still used. When nutrients are used up faster than they can be replaced, the soil becomes useless. It may turn to dust and blow away.

On a worldwide basis the average depth of usable topsoil is but seven inches. Upon these seven inches the people of the world depend for their nourishment. If soil erosion continues at the present rate, 7% of the Earth's soil will be used up each decade.[7]

Erosion is older than history. When early peoples overgrazed the land, erosion slowly but relentlessly destroyed the very land upon which they depended for livelihood and food. Aware of how erosion was adding to the problems of the depression years, Assistant Chief of the Soil Conservation Service W. C. Lowdermilk studied the record left by civilizations more than 7,000 years ago. He discovered that soil erosion was a major component in the decline of once-mighty civilizations. He found, for example, why the site of one of the wonders of the ancient world, the Hanging Gardens of Babylon, was now the scene of desolation. The area was made fertile by the waters of the Tigris and Euphrates rivers and a network of irrigation canals, but the ancient Babylonians were unable to stop the canals and rivers from choking up with silt (sand or sediment) that had washed off the lands being cultivated.

In case after case, Lowdermilk concluded that mismanagement of the soil was an important contributor to the decline of civilizations. He closed his report, *Conquest of the Land Through Seven Thousand Years,* with an "Eleventh Commandment" that read:

> Thou shalt inherit the Holy Earth as a faithful steward, conserving its resources and productivity from generation to generation. Thou shalt safeguard thy fields from soil erosion, thy living waters from drying up, thy forests from desolation, and protect thy hills from overgrazing by thy herds, that thy descendants may have abundance forever. If any shall fail in this stewardship of the land, thy fruitful fields shall become sterile stony ground and wasting gullies, and thy descendants shall decrease and live in poverty or perish off the face of the earth.[8]

Lest we go the way of these dead civilizations, this commandment needs to be observed. But protecting the land involves protecting its water resources, as well. This is the subject of the next chapter.

Notes

Citations in the notes are brief. Full citations appear in the bibliography.

CHAPTER TWO

1. *New York Times,* April 23, 1970.
2. Commoner, 32.
3. Ibid.
4. *New York Times,* November 14, 1988.
5. *New York Times,* April 21, 1986.
6. *New York Times,* September 17, 1988.
7. Brown and Wolf, 22.
8. U.S. Department of Agriculture.

C H A P T E R   T H R E E

Don't Go Near the Water: *Water Pollution*

In the summer of 1988, the advice "don't go near the water" had to be taken seriously. Many who tried to go swimming along the beaches of the Atlantic Ocean, Lake Erie or Lake Michigan were ordered to stay out of the water, which had been contaminated with hospital waste. Dirty bandages, vials filled with stale blood, and used hypodermic needles had all been dumped into the water. Raw sewage was another substantial contributor to ocean pollution and a dire threat to health. Worried that swimming in these waters might expose them to diseases such as hepatitis, or Acquired Immune Deficiency Syndrome (AIDS), most people faithfully heeded the advice. And so, during a summer that turned out to be one of the hottest on record, young and old alike did not dare go into the water.

Water: The Source of Life

Planet Earth is inappropriately named. There is so much more water on our planet than earth, that a better name for the globe on which we live would be "Water." "Adam's ale" has neither color nor taste, yet no living thing can live without it. Water is the common denominator of life, and of civilization. The civilization of ancient Egypt grew and flourished around the Nile River. The source of life for ancient Babylonia was the water of the Tigris and Euphrates rivers. India owed its ancient glory to the Indus River, while the ancient Chinese depended on the waters of the Yangtze River.

The spectacular development of the United States, no less than that of these ancient civilizations, rests upon its lakes, rivers and oceans. New York City owes its greatness in large measure to the Hudson River, at whose mouth it is located. Chicago could not have become one of America's greatest cities had it not been for its location on the shores of Lake Michigan. Cleveland became an important urban area because it is a port of entry on Lake Erie. And, where the Allegheny and the Monongahela rivers meet to form the mighty Ohio River, lies the city of Pittsburgh, once the steel capital of the United States and still a major industrial center. Swiftly running rivers throughout the country have been harnessed by means of enormous dams to spin giant turbines that generate electricity. Hydroelectric power, as this is called, reduces the nation's reliance on fossil fuels such as natural gas, oil and coal.

Securing control of the seas has been the goal of every nation that has sought world power. By controlling the routes of ocean commerce, a nation can enhance its influence in the international community. The seas remain essential commercial highways even as they have become a watery frontier of exploration as exciting as outer space. As we plumb the ocean depths (average: 12,500 feet), we increasingly look upon the sea as a source of additional food and minerals. From ocean-based drilling platforms, petroleum is extracted from deposits deep underwater in such areas as the Gulf of Mexico and Europe's North Sea. In huge oceangoing tankers, petroleum is distributed throughout the world.

The banks of New England, Nova Scotia, Newfoundland, Scandinavia and Japan are important sources of fish, an increasingly important part of our diets. Moreover, if it should become economically feasible to desalinate (remove the salt from) the ocean, an important source of fresh water would be available to a world that may soon need it.

Water may be thought of as a commodity like oil, coal, metals or agricultural products. It is needed in vast quantities, but unlike other commodities it is not generally bought or sold in the global marketplace. Vital to everything we do, water is rarely transported great distances. As a result, we have not set the price of water, as we have set the price of oil or pork bellies. There is a great uncertainty as to how much water is used, by whom and for what purposes. Priceless though it is, pure and plentiful water has been taken for granted. But this is a luxury we can no longer afford. We need to examine some of the problems of assuring ourselves sufficient amounts of good water.

Troubled Waters

Despite water's importance to America's development, most of the nation's streams, rivers and lakes are polluted. If, in days past, the lakes and rivers had been as filthy as they are today, great cities and industrial centers could never have taken root and developed. Now, maintaining a supply of fresh water for drinking and bathing is a problem for many American cities.

"Every time you take a glass of water from a faucet in St. Louis," reported the *Globe-Democrat*, the city's morning newspaper, "you are drinking from every flush toilet from here to Minnesota."[1] The Potomac River has been described as "an open cesspool."[2] Lake Erie was nearly a dead sea before local efforts restored some life to that body of water. In 1970, the Cuyahoga River, which meanders through Cleveland, Ohio, was so polluted with oily industrial residue that its surface spontaneously ignited periodically. In 1977, in Oswego, New York, toxic waste was discovered just before it entered Lake Ontario. A company whose business it was to dispose of industrial chemical wastes had stored the poisons in 115,000 corroding, leaking drums within the city. It had furthermore dumped the waste in a lagoon that ran into Wine Creek. From there the toxic chemicals made their way toward Lake Ontario. A local resident noticed the oil slick and informed the Coast Guard, which notified the Environmental Protection Agency.

The Hudson River is so polluted that it can no longer furnish water to the cities on its banks. Folksinger Pete Seeger wrote his "Hudson River Song" to publicize the current role of that once-mighty waterway as a filthy sewer. Swiftly flowing water has an impressive capacity to cleanse itself of moderate deposits of sewage, and plant, animal, human and industrial waste. But when the amount of waste dumped into a waterway becomes too high, that body of water becomes so overloaded with pollutants that it can no longer cleanse itself. A great deal of that sewage finds its way to the coasts, making many once-beautiful beaches unfit for swimming.

Polluted water cannot be used to water plants or irrigate fields. Fish living in the water die or become contaminated. Polluted waters give off odors that make the shores and the banks of rivers uninhabitable. Recreation opportunities such as fishing and swimming decline or disappear. Water shortages develop, touched off by a relatively small drop in the normal amount of rainfall.

Thirsty?

That most Americans have good drinking water is one of the engineering marvels of the American landscape. Out of sight, and therefore out of mind, a vast and sprawling underground network of pipes carries water great distances. Complex systems treat raw sewage, and sophisticated filtering techniques and chemical applications make our water safe to drink. Among the great, if seldom appreciated, wonders of the modern world is that so much drinkable water can be brought to great cities as well as to remote hamlets. While approximately four million rural poor lack water fit to drink, many millions of Americans have excellent drinking water. It is a remarkable achievement, but one that is now threatened.

On May 26, 1965, physicians in Riverside, California, reported treating an unusual number of stomach ailments. A thousand people had sought medical care for symptoms that included stomach cramps, diarrhea and bloody stools. A vast investigation was undertaken. No one suspected that the source of the illness was not in anything the victims had eaten, but in the community's water supply, which had been contaminated with fecal matter. The standard tests used in determining the purity of the water had failed to reveal the presence of any illness-causing bacteria. Chlorinating the water seemed to end the problem, but before the epidemic was over, 16,000 people had become ill, 170 were hospitalized, and three had died.

In 1969, with eight games left in the season, the Holy Cross College football team of Worcester, Massachusetts, experienced a plague. After a Saturday game with Dartmouth College, 20 players became ill with infectious hepatitis. Blood tests showed that 55 other players were infected, as were seven coaches, four managers, the trainers and the college's sports information director. Homecoming was cancelled and Holy Cross failed to win another game for the remainder of the 1969 season. The source of the illness was traced to the contaminated water supply on the Holy Cross football field.[3]

While some progress has been made, to date it has been halting. Several facts explain the less than impressive results. The first is that Americans have not yet recognized the importance of taking action to improve the quality of their water. The second is that such action is very difficult and very costly. The third is that the purity of drinking water is directly related to the source from which the water is drawn. These sources have become increasingly polluted and distant from the water-consuming community.

Barry Commoner drew a parallel between the current situation and an episode in the 19th century. In the 1870s, typhoid, a dread disease brought about by water polluted by human and animal waste, was a major health problem in this country. The cause and effect were self-evident. If typhoid was to be eliminated, fecal matter had to be removed from the rivers and streams that provided drinking water. Communities throughout the country were prepared to tax themselves in order to improve sewers and take whatever steps were needed to assure themselves a pure supply of water.

What the fear of typhoid was a century ago, the fear of cancer is today: one of the world's major environmental concerns. Evidence continues to mount that the dependency of agriculture and industry on chemicals has had residual effects that were not anticipated when those chemicals were introduced. Some of these effects could be deadly. Even chlorine—which at one time was heavily relied upon to maintain a safe supply of drinking water—is now considered a carcinogen if not carefully controlled. However, cause and effect are not always readily known. Often cancer will not appear for 20 years after exposure to a new chemical agent. Indeed, one can never really be sure that the cancer was directly caused by a specific pesticide or chemical, because other factors may have been at work.

Moreover, the source of water pollution may be distant from the community in which disease becomes evident. Wastes may originate in industrial plants located far upstream. Since the source of the pollution often lies outside the community affected, the question of which town should levy taxes to pay for the necessary cleanup is problematic. Industry, mindful of its balance sheet and uncertain as to whether it is indeed the villain, is often unwilling to invest in the complicated and costly equipment needed to maintain a clean supply of water.

And pollutants absorbed into the beds of lakes and rivers only intensify the problem. Dissolved as they are in this groundwater, they may have no odor and be undetectable. But when polluted, groundwater is a great danger. For one thing, it cannot be readily treated; for another, it does not merely settle at the bottom of the lake or riverbed. Instead, groundwater moves slowly and steadily, often to appear once again in an entirely different place. It may rise, for instance, into a new stream and poison that as well. Since all water at one time or another becomes groundwater, water pollution in one location can lead to polluted water far away.[4]

The Great Lakes Cleanup

The Great Lakes (Ontario, Erie, Huron, Michigan and Superior) stretch more than 800 miles from end to end, forming part of the border between the United States and Canada. (Only Lake Michigan is entirely within the United States.) They extend over 95,000 square miles, containing 20 percent of the Earth's fresh water, and 95 percent of the surface fresh water found in the United States.[5]

Sad to say, however, the water in the Great Lakes is no longer sweet. A 1985 study conducted by the Royal Society of Canada and the U.S. National Research Council showed that the population of the Great Lakes region is exposed to a higher concentration of chemical toxins than any other population of North America.[6] That these people should have health problems is not so surprising given the built-up industrial landscape and polluted waters of these lakes at present.

Among the characteristics common to the lakes is that it takes a long time for a drop of water entering a lake to leave it again.

In a way, lakes cleanse themselves, replacing water lost through evaporation and seepage with fresh rainwater. How long it takes for a drop of polluted water to leave a lake, therefore, depends on how large and how deep the lake is. For instance, Lake Superior, the largest and deepest of the Great Lakes, takes 200 years to renew itself, while Lake Erie, shallow and smaller, takes only two years. The long residency rate of a drop of water and the overdevelopment of the Great Lakes region explain the persistence of over 900 known or suspected toxic substances in the water there. Little wonder then that the Great Lakes have been described as the "world's largest sewer."

In the 1960s, Lake Erie was choking to death because of the inadequately treated sewage being dumped there by the cities of Detroit, Cleveland, Buffalo and Toledo. So thickly was it polluted with green algae, people joked that you could walk on Lake Erie. The stench from the waters was terrible. Millions of dead fish and foam from detergents washed ashore on Lake Erie's beaches. None of the other lakes was as badly polluted, but all had substantial pollution problems, which sometimes seemed beyond solving.

Sewage is not the worst pollutant in the Great Lakes, however. They are also tainted with dangerous chemical compounds capable of causing cancer and birth defects, such as PCBs, toxaphene, dioxins, and DDT. Investigators found the source of these toxic chemicals in the water runoff from farms that had used them as insecticides or pesticides. Leaky toxic-waste sites and runoffs from urban streets were other sources of the contaminants.

To the surprise of those trying to figure out why the Great Lakes seemed to be a magnet for toxic wastes, they learned that many of the toxins had been airborne and transported to the region by winds. The expanse of the Great Lakes' surface area provides an ideal landing area for airborne pollutants. As a result, those who live near the Great Lakes are exposed to danger on many levels: in the air they breathe, in the water they drink and use for recreational purposes, and in the fish they eat from the Great Lakes.

Toxic wastes find their way into the fatty tissue of fish and cause skin lesions, lip sores and cancer in the fish. If the fish are cancerous, are lives of those who eat the fish in jeopardy? If so, how much of a threat are these fish, and what can be done about it? Pregnant women are advised to avoid eating some of the fish of the Great Lakes. A group of over 500 men and women who live around the Great Lakes and normally eat many of the fish found there are currently being studied. The outcome of these studies will be important to all of us. The Great Lakes are central to the well-being of the nation. In the war against environmental pollution, the Great Lakes are surely a major battlefield.

But it is not only the well-being of the United States that is at issue. Canada, our neighbor to the north, likewise has a giant stake in what takes place on the Great Lakes. In 1978, the United States and Canada signed the Great Lakes Water Quality Agreement. This treaty was a forward-looking measure in that it addressed the problems of the "ecosystem" of the Great Lakes. It was the first international agreement concerning such an ecosystem. Forceful in language, clear in its intent to achieve the "virtual elimination" of toxic substances, the agreement gives direction to the two nations in their efforts to solve the problems of the Great Lakes.

Some progress has already been achieved. Lake Erie, for example, is no longer an environmental disaster. However, environmental dangers persist. The more difficult problems of how to dispose of toxic chemical compounds that are not easily dissipated has not been adequately addressed. Progress is slow, because neither nation can be forced to implement the agreement or to enforce its provisions with the same zeal expressed in the treaty.

Water for Tomorrow

Endowed with approximately 370 quintillion (a number with 18 zeros) gallons, the world has plenty of water.[7] Ancient people survived on about five gallons of water a day. But today, according to Thomas Canby in a *National Geographic* piece, each person uses about 87 gallons: 24 for flushing, 32 for bathing, laundry and dishwashing, and 25 for swimming pools and watering the lawn. Only two gallons are used for drinking and cooking. Water in great quantities irrigates our food and fiber and supplies our factories. It is astounding to realize the amount of water needed in the creation of things we take for granted: eggs require 120 gallons of water each; steak, 3,500 gallons; a ton of steel, about 60,000 gallons. Altogether, direct and indirect water use per person today is about 2,000 gallons.[8]

The amount of water needed for industry and agriculture will grow even more dramatically during the coming decades. Because so much of our water has become unfit for use, the nation may not have all the water it requires. And, since industry uses most of our water, our biggest problem will be meeting its growing needs.

The problem of water supply in the United States has traditionally been viewed regionally. Water quality has been the issue in the East, while water quantity has been the issue in the West. At one time there was some truth in this generalization. The overcrowded East, which relied upon a deteriorating system of reservoirs and aqueducts to supply aging cities, could not cope with the industrial wastes pouring out of its ever-growing factories. Recently, however, the availability of water has become a growing problem in a number of areas, such as Long Island in New York State.

The West, on the other hand, was populated by pioneers who believed that "rain follows the plow." They hoped to farm land in arid areas, but rain in the West was often disappointing. Sixty percent of the nation's land—the West—receives but 25 percent of its rainfall. Unfortunately, however, water scarcity is not the West's only problem. The West is rapidly using up its groundwater, and many of its water sources are becoming contaminated. For example, the pesticide DBCP (dibromopropane), a carcinogen, has been found in the groundwater of many communities of the San Joaquin Valley in California.

In a vivid scene in the movie *Chinatown*, the character played by Jack Nicholson got his nose brutally cut because he dared to intervene in a conflict over water rights in California. The rights to water for drinking, farming and manufacturing have been fighting issues in the East as well as the West. A classic example of the conflict may be found in the use of water from a vast underground reserve called the Ogallala Aquifer.

An aquifer is an underground geological formation that contains fresh water. The Ogallala Aquifer, as large as Lake Huron, provides water for eight states from South Dakota to Texas. Named after an Indian tribe, it is roughly three times as big as the state of New York. For thousands of years, there has been water in the Ogallala. It probably accumulated even before there were men and women to use and appreciate it.

Today this ancient and precious resource is being overused. Because the Ogallala Aquifer is located in an area of limited rainfall, little water seeps through the Earth to replace the vast amounts that are being withdrawn. Thus, more water is being taken out of the aquifer, mostly for irrigation, than is being returned. Hydrologists (water geologists) refer to the process as water mining. Today, a billion-dollar agricultural economy is based on the availability of water in the Ogallala Aquifer. When this water is all consumed, as it may soon be, what will happen to this enormous industry and to all who are dependent upon it?

The settlers who came to the American prairie in the late 19th century expected rain but found that drought was their more likely fate. Hamlin Garland, author of *A Son of the Middle Border*, describes how such a farm family was subject to the whims of nature. Of his father he wrote,

> All his life he had been menaced by the weather. Clouds, snows, winds, had been his unrelenting antagonists. Hardly an hour of his past had been free from the fear of disaster. The glare of the sun, the direction of the wind, the assembling of clouds at sunset—all the minute signs of change, of storm, of destruction had been his incessant minute study. For over fifty years he had been enslaved to the seasons.[9]

In 1902, in response to the need to help western farmers, the United States Congress created the Bureau of Reclamation. The mission of the bureau, now a division of the Department of the Interior, was "to reclaim the arid lands of the western United States for farming by providing a secure, year-round supply of water for irrigation." The legacy of the bureau was the system of great dams built in the mid-1930s, such as the Hoover, Shasta, Grand Coulee, Bonneville and Fort Peck dams—at the time, all among the world's largest. Noble achievements though they were, they did not solve the problem for good. During the summer of 1988, the prairie states experienced another drought. Farmers feared that it might continue for a number of years, as such droughts had in the past. Clearly, the great dams of the 1930s are no longer adequate for the water needs of the burgeoning West. But building more dams and bringing more water ever greater distances to drought areas is a costly and complex proposition.

One proposal would draw upon the waters of Hudson Bay in Canada. By damming the bay's lower portion and desalinating its water, many gallons of much-needed water could be delivered annually to the Colorado River. Another proposal would divert waters from Alaska and northwestern Canada to the United States and Mexico through a complex network of canals and reservoirs, including a 497-mile trench reservoir in the Rocky Mountains. But such a project, according to political scientist Thane Gustafson, quoted in *Environment* magazine, "could sooner be built on the moon than in the United States or Canada."[10]

Visionary schemes abound, some fanciful or even bizarre. Can icebergs be towed from Antarctica to the California coast so that water from melting ice can be pumped ashore for human consumption? Can the oceans be desalinated on a grand scale at reasonable cost?

There is really no water shortage, only water mismanagement. Water conservation would be relatively easy if Americans became more conscious of how they use water. If the demand for water could be cut in half, the availability of water would be doubled. "Conservation...can be practiced without pain or suffering or deprivation, even without sacrificing one's precious 'life style,'" wrote Fred Powledge in *The Nation*.[11] Here are some water-conserving techniques:

- A dime-sized metal disk with a hole in it, placed in the shower head, reduces the volume of water used in showering without altering the spray.
- Devices to reduce the amount of water used in flushing a toilet are readily available.
- Waste water captured from the shower into a bucket can be used to flush the toilet or feed houseplants.
- Using less hot water reduces the amount of energy that must be used and is a water saver as well.

Techniques are being developed to encourage the recycling of water used for irrigation and industry. However, such recycling requires that sewage be carefully treated to avoid a health hazard. Nevertheless, the technology is there. In the long run, such measures may be less expensive than moving icebergs or bringing water inordinate distances. One way or another, Americans need to reduce their use of water, and improve their management of water resources.

Notes

Citations in the notes are brief. Full citations appear in the bibliography.

CHAPTER THREE

1. Gordon, 84.
2. Herber, 81.
3. Sheaffer and Stevens, 55–57.
4. Carson, 47.
5. Cobb, 4–5.
6. Ibid.
7. Postel, 5.
8. Canby, 144–48.
9. Garland.
10. Sherman, 11.
11. Powledge, 718.

CHAPTER FOUR

Hold Your Breath: *Air Pollution*

During the 1960s, young people across the country were demonstrating against the Vietnam War and for civil rights. They also demonstrated about the quality of air in the major cities of the country. They were right to be concerned.

Airborne Dangers

Concern about the quality of air we breathe is not new. Air pollution began when humankind learned the benefits of fire. Smoke began to pollute the environment of primitive peoples. The citizens of ancient Rome complained that soot smudged their wool togas. England's King Edward I, who ruled from 1272 to 1307, had a man put to death for burning coal instead of oak; the monarch believed coal to be more polluting. When the Spanish conquistador Hernando Cortes and his men traveled to the New World in the 16th century, they saw a haze over the mountain-ringed basin that now contains the city of Mexico City. Indians there were polluting the air with the smoke of their campfires.

At the turn of the century, in American cities, people relied on the horse to transport the population and haul heavy loads. That versatile beast of burden deposited an average of 25 pounds of manure on the streets of the city every day. In hot weather, the manure attracted swarms of flies while traffic ground the dried manure to dust. The dust, in the words of a contemporary (quoted by columnist George Will in *Cape Cod Times*, September 19, 1988), "flew from the pavement as a sharp, piercing powder, to cover our clothes, ruin our furniture and blow up into our nostrils."

The solution to the manure problem was the invention of the internal-combustion engine, which was soon used to drive automobiles. But, as we know, this did not end the problem of air pollution. Indeed, the automobile has *aggravated* the problem. Here are a few examples:

- In 1948, a mysterious fog covered Donora, Pennsylvania. Thousands became sick, and 20 died. The rates of death and sickness remained high in that community for many years afterward.
- In 1952, in the city of London, England, there were 4,000 to 5,000 more deaths than normal. A cloud of stagnant dirt and soot had hung over the city for about two weeks.
- On Thanksgiving Day, 1966, New Yorkers learned that an inverted air mass had settled over the city and would not allow wind to dissipate the city's smoke and soot. Fortunately, the condition did not continue for very long, and a "smog alert" was avoided. Had a smog alert been declared, people with heart and breathing problems would have been advised to remain indoors. The use of private automobiles would have been prohibited except in special circumstances. Incinerators in apartment buildings would have had to shut down, and many factories would have had to cease operation.
- In the quiet California town of Casmalia (population: 200),

people demonstrated against a toxic waste dump where cyanide wastes, oil-field acids, and industrial solvents were dumped in open lagoons and left to evaporate. Local residents complained of nosebleeds, respiratory problems, chronic diarrhea, birth defects, and at least one death from leukemia.

Clearly, dirty air is not limited to the large urban centers of the East but is a national, indeed, an international problem.

Our very breath is a pollutant. We inhale oxygen and exhale the pollutant carbon dioxide. Moreover, lightning bolts, sea spray, volcanoes and trees annually release pollutants like nitrogen oxides, sulfur and hydrocarbons into the air.[2] Within limits, the air around us can absorb the pollutants (dirt, dust, smoke) that are fed into it. We are, therefore, safe as long as we do not release more pollutants into the atmosphere than can be eliminated naturally. It is when this balance is not maintained that our worries begin.

In a sense, air pollution is the price we have paid for progress, which has taken the form of large factories that employ many people to manufacture life's necessities and luxuries. Large cities intensify the problems of air pollution. The exhaust from our cars and buses makes the air unpleasant to breathe. The chimney soot and smoke from the furnaces that heat our homes and produce electricity contaminate the air. Chemical waste given off by factories adds new dangers to the environment.

Among the major causes of air pollution is the release of unburned fuel into the air, usually in the form of smoke or exhaust. Because fuels do not burn completely, a percentage of unburned fuel is released as exhaust into the air. Automobiles are one of the major causes of this type of pollution, much of which escapes through the exhaust system into the air as carbon monoxide, a deadly killer.

Smoke, dust and fumes—all soil the air we breathe. Factories that convert petroleum into gasoline emit a smoke so filled with acid that it makes one's eyes and throat burn. Jet planes release a poisonous gas that pervades every major airport of the land. It is little wonder that few cities wish to expand their airports or build new ones.

Air Pollution Report Card

For both achievement and effort in the fight against air pollution, the nation's performance hasn't been impressive. Dennis Hayes, writing in a World Watch Institute publication, notes that "the principal pollution-control efforts have gone into solving those problems that appeared most solvable. Devices have been attached to automobile exhaust pipes, to industrial chimneys, to sewage pipes—all in an effort to remove certain pollutants. The pollutants thus removed have not always been the most dangerous, but they have been the most 'removable.' The 'tonnage' of pollution has thus decreased dramatically....But much of the danger remains."[3]

Despite the requirements of the Clean Air Act of 1970, America's largest cities—New York, Los Angeles, Chicago, Philadelphia, Detroit, Washington, Houston and Boston—still have too much smog. Smog consists mainly of ozone, which can cause serious problems in breathing. In the upper atmosphere, however, ozone is desirable, because it helps to block the sun's potentially cancer-causing ultraviolet rays. But upper-level ozone is being destroyed by chlorofluorocarbons, the gases used in air conditioners, refrigerators, aerosol sprays, and in the manufacture of plastic foam.

The Clean Air Act of 1970 identified sulfur dioxide, nitrogen oxide, carbon monoxide, ozone, lead and solid particles as the nation's most pressing air-pollution problems. Because studies in 1980 showed that most Americans had excessive amounts of lead in their blood, and because some learning problems were linked to lead poisoning, lead was removed as an octane booster in gasoline. Factories were required to control lead emissions as well, and soon the air above industrial cities was largely cleansed of soot. The nation has been less successful in removing ozone, carbon monoxide and nitrogen oxide from the air.

Hidden Dangers

One's home may be a trap for pollution that is neither readily detected nor easily prevented or corrected. Indoor pollution has many victims but few obvious perpetrators. The outdoor air we breathe may now be a bit cleaner than it once was, but the indoor air can still pose serious health hazards. Here are some examples:

- Tobacco smoke can cause lung cancer and other respiratory ailments. Those who smoke expose not only themselves but also others to risk in their homes.
- Asbestos, found in floor, ceiling and pipe insulation, can produce lung disease and intestinal cancer. The National Education Association branded asbestos the "silent killer," and the Environmental Protection Agency banned the use of asbestos in certain building materials. Many communities have begun massive programs to remove asbestos from schools.
- Formaldehyde, found in plywood, particle board, and foam insulation, causes cancer in animals and irritates eyes, skin and lungs.
- Benzine, widely used in dry cleaning, may cause leukemia.
- Styrene, found in carpets and various plastic products, may damage the kidney and liver.
- Radon has been identified by the EPA as a particularly dangerous gas. It is a colorless, odorless gas released by

decaying uranium, which is found in various levels of concentration in soil and rocks. After years of exposure in their homes, victims of radon may develop lung cancer. In 1988, the EPA reported that about eight million American homes had dangerous levels of radon and issued a number of bulletins advising what to do about the gas. Worried homeowners and those who live in apartments located on lower floors can use simple kits sold in supermarkets to test their homes for radon levels. Should high levels of radon be found, improving ventilation and sealing cracks in the home's foundation can help.

Improving air quality in the homes in which we live and in the factories and office buildings in which we work must take a higher priority in the future. The solutions to internal air pollution are not complicated, but they often require a change in lifestyle. The problems of external air pollution also require our attention. Despite some progress, they continue to grow ever more complex.

Love Canal

Niagara Falls, New York, is a tourist attraction, a natural wonder, an industrial complex, and a popular site for honeymooning couples. To the east of this romantic spot lies an abandoned community known as Love Canal. The town has become a metaphor for what could happen to the entire American landscape, should we fail to begin caring properly for it. Those who once lived at Love Canal were forced to abandon hope.

Love Canal was destroyed by waste from chemical plants. These companies had legally dumped chemicals in the abandoned canal since 1947. Love Canal became "...our country's baptism into the problem of toxic chemical waste," according to David Weinberg in an article for *Natural History* magazine.[4]

In 1953, officials of the city of Niagara Falls insisted that Hooker Chemical sell the 16-acre dump site to the city or have it confiscated. Hooker transferred the land to the city's board of education for one dollar and warned the community about the poisonous wastes that were in the canal. Nevertheless, an elementary school was built on the site and some of the land was sold to real-estate developers. A few years after development began, problems became evident. As one resident quoted in *Audubon* magazine described it, "The area was smelly. The trees near the vacant lot would lose their bark, and leaves were falling year-round."[5] A reporter for the *Niagara Gazette* had some of the water analyzed. A laboratory determined that the water contained excessive levels of 82 industrial chemicals, seven of which were suspected carcinogens.

In addition, a health survey determined that there were high rates of miscarriages, birth defects, and liver and lymphatic problems in the local population. Nearly half the residents had high white-blood-cell counts. Finally, 239 homeowners closest to the canal were told to seek other housing immediately. A total of 423 homes were eventually abandoned.[6]

In one sense, Love Canal is not unique. It is but one of thousands of communities endangered by toxic waste dumping. The town has been partially cleaned up, and some former residents are making a cautious return. Some new people are even moving into the area. But as difficult as it is to deal with pollution arising from toxic waste dumping, an even more complicated threat to the environment is acid rain.

Acid Rain

Acid rain is a legacy of the Industrial Revolution. When sulfur and nitrogen oxides, formed when fossil fuels (coal, oil, natural gas) are burned, are carried into the atmosphere, they become acidic. They are eventually dissolved in atmospheric humidity and return to Earth as acid rain. Public utilities, factories, homes, automobiles and industrial processes contribute to acid rain. The term acid rain includes snow, fog, mist, dew, hail and all other matter that is absorbed into the atmosphere and returns to Earth as precipitation. Acid rain can be dry as well as wet. All rain is slightly acidic, but the term acid rain applies to either wet or dry precipitation with high concentrations of nitrogen or sulfur dioxide.

Acid rain weakens bridges, damages buildings, destroys plants and forests, and kills animals and fish. Lakes and drinking water are contaminated by acid rain. And it is already responsible for forest destruction in the Great Smoky Mountains and in New York's Adirondacks and Catskills. In lakes contaminated by acid rain, fish become diseased and can no longer reproduce, and fish-eating birds consume contaminated fish and die. Insects and other foods that smaller birds find along the shores of the lakes are likewise poisoned, and they too die or move elsewhere. Many experts believe that acid rain has become a major health hazard, and they rank it third, behind active and passive smoking, as a cause of lung disease.[7]

Because acid rain corrodes the steel in bridges, constant vigilance is required to ensure that these structures remain safe. The great architectural treasures of the world, such as the Parthenon in Athens, the Taj Mahal in Agra, India, the Colosseum in Rome, Westminster Abbey in London, the cathedral of Cologne, Germany, and the Capitol building and many of the monuments in Washington, D.C., are all in danger. They may be totally destroyed or damaged beyond recognition unless measures are taken to reduce the effects of acid rain.

Not everyone agrees that acid rain is such a grave problem. Some claim that it is hard to find a cause-and-effect relationship between the deterioration of architectural treasures, pristine lakes and huge forests—and the burning of fossil fuels. Scientists are researching, but it may take a scientific miracle to find the exact answer.

Part of the problem is that the source of acid is often far from the area it affects. The nine largest coal-burning states, for example, are in the Midwest. Tall smokestacks, originally designed to control local pollution, spew sulfur and nitrogen oxides into the atmosphere. The chemicals are carried eastward by prevailing winds; they destroy lakes and forests in New England and Canada. The same smokestacks installed to reduce environmental pollution may thus be a major factor contributing to acid rain. Some acid rain is also produced in Canada, and may contribute to environmental destruction in the United States. The consensus is, however, that most of the acid rain now destroying parts of the Canadian environment is produced in the United States.

Because the effects of acid rain turn up many miles from the source, paying for pollution control is a confusing problem. If acid rain is indeed the culprit in a substantial amount of environmental pollution, it will require about $50 billion to fit factories with cleansing mechanisms such as flue-gas scrubbers. If emissions of sulfur and nitrogen dioxides are to be reduced, there will have to be quite a change in the lifestyle of Americans. Is this a price Americans are willing to pay? That is, will they drive less, conserve more, and use fewer power tools? Are Americans willing to pay if there is some lingering doubt that acid rain is a real enemy? Moreover, with technology changing so rapidly, in which technology should we put our trust?

Further, should the states in which the acid rain originates pay for cleaning up states many miles away? Or should the states that benefit from the cleanup bear the financial burden? Should the cost of curbing acid raid be shared by all Americans? And since acid rain, like many other environmental problems, is a global problem, how should the world community be involved? These questions indicate the intensely political nature of the debate on reducing acid rain.

Acid rain is an environmental time bomb. We do not know how long it takes for acid to build up enough to damage the environment or to threaten human health. We do know that acid rain is produced from sources as varied as volcanoes and the burning of fossil fuels. We can do little about the former, but to the extent we can control the latter and thereby improve our environment, we must address the problem of acid rain without further delay. Unless we control the controllable, acid rain may continue to contribute to another environmental problem: the long-term warming of the Earth's climate. This trend, called the greenhouse effect, may alter the very conditions of life on Earth.

Greenhouse Effect

Within your lifetime, your home on planet Earth may become a warmer place. The "greenhouse effect," as this is called, is caused in part by the burning of fossil fuels such as oil, coal and natural gas. Carbon dioxide, methane gas, nitrous oxide and chlorofluorocarbons are released into the atmosphere during fossil-fuel combustion. Once released, they destroy the protective shield known as the ozone layer. These emissions are all by-products of the way we live, including how we heat our homes, the kind of cars we drive, and the industrial processes we use to manufacture and package our goods. The styrofoam cups and packages used by fast-food restaurants are made with blowing agents. These agents are chlorofluorocarbons, a group of chemicals that contribute to the destruction of the ozone layer.

The ozone layer makes it possible for us to enjoy the sun, because it protects us from the sun's full strength. When the ozone layer becomes weak, more of the sun's damaging rays reach the Earth's surface. These rays may cause eye disease (cataracts) and skin cancer. A weak ozone layer allows the sun's more dangerous rays to enter the atmosphere, just as the glass in a greenhouse allows the sun's heat and light to enter. But, as in a greenhouse, the warmth of the sun is trapped inside. It has no way to escape. When the ozone layer is decreased the planet warms.

According to a *World Watch Institute Report*, the scientific community has reached a consensus that "...the projected warming appears to have begun....Many scientists now believe that important climate shifts could occur abruptly, causing losses in agricultural production that could not be readily adjusted for....This global warming could occur when children born this year [1988] are in their forties."[8] The 1980s were the warmest years on record in a century, and several summers were extraordinarily hot. The level of carbon dioxide in the atmosphere is likely to be 60% higher by the mid-21st century. A rise in the Earth's temperature seems inevitable.

You might think that this may not be too bad, especially if you live in the upper Midwest or New England. Such a warming, however, could bring dramatic changes, most of them undesirable. If the climate on planet Earth warms up even a little, the American landscape could change in a number of ways:

- The sea level could rise eight to sixteen inches, endangering coastal areas. A state like Massachusetts could lose about 10,000 acres of land to the sea.
- Farmers in the Midwest could experience even hotter and drier summers. During the summer of 1988, one of the hottest on record, a drought in the Midwest drained the Mississippi River to the lowest point ever measured. Some feared that the summer of '88 was a harbinger of more droughts to come. When the anticipated greenhouse effect

arrives, land now producing wheat will produce only grass. Rich corn land will be devoted to crops that need less water, such as wheat. The economic consequences of such shifts in land productivity are difficult to estimate, but the change would clearly be costly. Climate shifts could also force population shifts.

- Water levels in the Colorado River could drop, disrupting farming, water supplies and electricity generation in eight states, including California. Rainfall patterns would change, resulting in a need for costly new systems of irrigation.

Buying Time

Governments worldwide are trying to make plans for coping with the greenhouse effect. Many nations realize that shifts in agricultural productivity, water availability and transportation routes could change the international balance of power. Private industry, too—especially utilities that burn a great deal of fossil fuel; mining, lumbering and oil-drilling companies; and food processors—are likewise studying what they can do to offset the ill effects of global warming. Meanwhile, nations and industry are trying to slow down, if not stop, the global warming trend. But whatever we attempt to do will be costly, and will require changes in lifestyle and living standards.

Natural forces contribute to the greenhouse effect, but it is our use of fossil fuels that is speeding the warming of the planet. An obvious—but not necessarily the best—solution is to reduce our use of fossil fuels. But the cost would be immense, requiring huge reorganizations of industry. Coal and oil contribute more carbon dioxide than natural gas, but even shifting to the latter creates problems. For example, the United States has an estimated 300-year supply of coal, but only a few decades' worth of natural gas. If we switch to natural gas and then find that it is used up in one or two generations, what then? Moreover, using natural gas is more costly than oil and far more costly than using coal.

In an attempt to produce cleaner electricity, utilities have recently installed equipment to reduce the amount of sulfur dioxide that they release into the environment—a major contributor to acid rain. Similar attempts to reduce emissions of carbon dioxide can be achieved only at enormous expense. Were the utilities to undertake this, in many cases, it would cost more than the original plant. To recover their costs, utilities would have to increase electric rates by 75%. Should consumers pay? Would consumers pay?

The United States consumes more energy per capita than do comparable industrial nations such as Germany and Japan. If federal standards for car and truck mileage were to go up, there would be a further reduction in carbon dioxide emissions. But in 1988, when the federal standard was about to go up to 27.5 miles per gallon for cars, the Reagan administration, at the urging of automobile manufacturers, lowered the standard to 26 miles per gallon. Gas guzzlers have increased in popularity as gas shortages no longer seem to threaten. Would you be content with a smaller and less powerful car? American consumers prefer high performance to high efficiency in their automobiles, and Detroit is eager to exploit this preference.

Fast-food outlets use styrofoam cups and similar types of packaging manufactured with the use of chlorofluorocarbons, the chemicals responsible for the weakening of the ozone layer. In order to halt further deterioration of the ozone layer, the EPA has set a goal of reducing by 60% the amount of ozone-depleting chlorofluorocarbons we use. Refrigerators, air conditioners, plastic foams, cleansing agents and food refrigerants are among the leading products that require chlorofluorocarbons. A reduction in their use will have a sweeping impact on industry and on consumers. Consumer goods would become more expensive. Workers employed in the industries that manufacture and sell these widely used products would have to find other jobs. "Virtually every American household," says Douglas G. Cogan of the Investor Responsibility Research Center in Washington, D.C., "most of the nation's transportation fleet, and 375,000 business locations will experience withdrawal symptoms as the nation weans itself of its daily dependence on these chemicals."[9]

The America of the early settlers was one of forests. Today, these forests are gone. Not only have many of the forests disappeared from North America, they have also disappeared from many other places on the globe. Their destruction has caused a shortage of wood and reduced nature's ability to rid the environment of poisonous carbon dioxide. In the process of photosynthesis, trees absorb carbon dioxide and release oxygen into the atmosphere.

Deforestation reduces the number of trees performing this vital function, thereby contributing to carbon-dioxide pollution and the greenhouse effect. A program of reforesting the Earth could thus help delay the Earth's warming.

The greenhouse effect is not limited to America—it is a global problem requiring global collaboration. This is difficult to achieve. Brazil, for example, has embarked on a massive program of cutting down large portions of the Amazon rainforest in order to make room for cattle ranches. The Amazon rainforest is more important than any other single forest in reducing the amount of carbon dioxide in the Earth's atmosphere. What Brazil is doing affects everyone on Earth. Some nations may even feel that they stand to benefit from a warming trend if it makes unproductive lands agriculturally rich. Why should they try to delay the greenhouse effect? Why should their people make sacrifices to reduce the emissions of carbon dioxide, methane and other gases that contribute to global warming? Environmental problems increasingly require global responses.

If the greenhouse effect is to be addressed in the United States, a careful analysis must be made of our energy policy. Energy is a vital component of everything Americans do. Because our energy policy is so central to the state of the American environment, we turn now to an examination of our energy use and some of our alternatives.

Notes

Citations in the notes are brief. Full citations appear in the bibliography.

CHAPTER FOUR

1. Will.
2. Grove, 508–09.
3. Hayes, 5.
4. Weinberg, 12.
5. Hanson, 14.
6. Hanson, 14–15.
7. *International Wildlife*, "Clean Air, Maybe," January/February 1988, 26.
8. Brown et al., 16.
9. *Futurist*, "Economics of Preserving the Ozone Layer," January/February 1989, 41.

CHAPTER FIVE

Energy: *The End of the Free Lunch*

Because humans are born without fur or feathers, men and women rely upon numerous sources of energy in order to clothe, feed and house themselves. The history of civilization is reflected in the many forms of energy people have used to make clothing for warmth, cook food for nourishment, heat their homes for comfort and move about from one place to another.

Early men and women harnessed the energy of the sun, fire (wood), wind and water. In modern times people have used the fossil fuels of coal, oil and natural gas. We have now entered the age of nuclear energy, but because of its dangers we are once again considering how we can use ancient sources of energy, especially the sun, wind and water, for clean, safe, cheap energy.

Energy from Fossil Fuels

Coal, oil and natural gas are the major fossil fuels we depend upon for energy. They are called fossil fuels because they are formed from plant life that died eons ago. When fossil fuels are burned, they provide energy in the form of heat, which is then used to produce other forms of energy. Fossil fuels have two major common characteristics: when burned, they are almost entirely used up; and they do not burn cleanly, but pollute the environment into which they are released.

There is no shortage of fossil fuels, yet because we are using them up faster than they can be created (fossil fuels take hundreds of millions of years to form), there is great concern that we may not have enough in years to come. The search for new sources of fossil fuel continues at an ever accelerating rate, but most scientists believe that the world is unlikely to identify any major new deposits.

The rate at which fossil fuels are consumed can be considerably slowed down through conservation, making our supply last much longer. For example, in 1987 the United States used 56 tons of oil or its equivalent in coal and natural gas for every $1,000 of goods and services produced. Switzerland achieved the same rate of production with only 34 tons, and Japan with 28 tons, according to the International Energy Agency. There is, it would seem, ample room for more efficient use of energy in the United States.

Since fossil fuel sources are not equally distributed among the nations of the world, those who control important fossil fuel deposits are in a position to gain great wealth and power. Conflict over the control of fossil fuel resources, especially oil, can trigger regional and even global conflict. Those nations that have fossil fuels can use them as a weapon against nations that need them. In 1973, the United States launched what it called Project Independence, to make America independent of foreign sources of energy, particularly oil. Today, however, our dependence on foreign sources continues unabated.

Fossil fuels are not clean sources of energy. When they burn, they pollute the environment with poisonous gases such as sulfur, carbon dioxide and other dangerous chemicals. Given the self-cleansing power of the environment, some of the pollution caused by burning fossil fuels can be dissipated harmlessly into the atmosphere. But as a result of the increased use of fossil fuel, noxious gases linger in the air creating complex problems, such as acid rain and the greenhouse effect. For these reasons, alternatives to fossil fuels are needed, but old habits die hard. We are reluctant to give up the luxurious lifestyle based upon the lavish use of these fossil fuels. "Yet the time when the era of free lunches, which fossil fuels have afforded us, comes to an end is almost upon us," writes Don Hedley in *World Energy: The Facts and the Future.*

As we move into the next century, we will continue to rely on fossil fuels to meet our needs. But the world's supply of fossil fuel is finite, and someday we will run out. Burning these fuels contributes to environmental destruction, and alternatives need to be found. However, whether we concentrate on energy conservation, find alternatives in the form of renewable resources, or expand our reliance on nuclear energy, we will find no quick fixes. We have sent astronauts to the moon, but that may have been easier than finding effective energy alternatives.

Old King Coal

As a safety measure, coal miners used to take canaries into the mines with them. When the air contained too little oxygen, the canaries died, and miners knew it was time to stop mining and move elsewhere. Today, there is a grave concern that not only canaries but entire species of birds and animals—as well as people—are all in danger because of the poor quality of our air. The burning of coal is a major factor in air pollution. Yet the utilities that supply us with electricity all rely heavily upon coal, the most plentiful source of energy we have.

The United States has sometimes been called the "Saudi Arabia of Coal," because it has so much of it. In 1758, 32 tons of coal were shipped from Virginia to New York in the first recorded coal shipment in America. In 1770, George Washington was so eager to examine the potential of coal that he paid a visit to a colonial coal mine. Until 1880, however, wood was the most widely used fuel in the United States. By 1925, the United States used coal for 70% of its energy. But, following World War II, oil and gas surpassed coal as an energy source because they were more easily transported and were thought to be less polluting.

Coal deserves its reputation as a polluter, but it also deserves credit as a source of America's industrial might. Although more than 50% of America's electrical utility companies use coal, it also has many drawbacks. In addition to those mentioned above, coal mining is a highly dangerous occupation, and loss of life from accident and disease is substantial. East of the Mississippi River, coal is drawn from mines deep underground, but most of it is taken from surface or "strip" mines by giant steam shovels. In both cases, but more often in strip-mining than in deep mining, the removal of coal causes horrible scarring of the land. In recent years, coal- mining companies have gone to great lengths to try to restore mined land to its original productivity and beauty.

Despite grave problems, giving up coal altogether is more easily said than done. Because it wants to continue to take advantage of coal's availability, the challenge to industry has been how to make the use of coal less polluting. There are a number of promising new technologies that may make it possible to burn coal more efficiently and cleanly. One such technology adds limestone to the furnaces in which the coal is burned. The limestone particles react with the sulfur produced by burning coal, removing up to 95% of the impurities.[2]

In one California power plant, the nitrogen and sulfur in coal are converted into a synthetic gas (syngas). The gas is burned efficiently to spin turbines, which in turn generate electricity without the creation of major pollutants. The significance of this process is that a single plant both eliminates the problem of toxic coal wastes and efficiently generates electricity.

In order to reduce the costs of coal transportation, attempts are being made to pulverize coal into slurry coal, which can be sent through pipelines to wherever it is needed. The railroad industry does not support this technique, because it will lose the revenue and profit derived from transporting coal.

Coal is an old energy source with an uncertain future. However, it is clear that for many years to come, the nation will continue to use large quantities of coal. With oil and natural gas in limited supply, coal's very availability means that it can help the nation meet its energy needs. This use of coal will help us reserve our supply of oil and gas for purposes for which coal cannot be substituted. Until new techniques of generating energy can be developed or renewable energy sources identified, coal will remain the bridge to the nation's energy future.

Oil: Liquid Gold

In 1950, the price of gasoline at the pump was 27 cents a gallon, of which 20 cents were for the gasoline and 7 cents for taxes. By early 1973, the price of gasoline had risen moderately to 37 cents, of which 11 cents were for taxes and 26 cents for the gasoline.[3] During 1973, however, American cars screeched to a halt in the face of a gasoline shortage. The nation learned a painful lesson about the unreliability of the sources upon which it depended for oil.

That year, the Organization of Petroleum Exporting Countries (OPEC, consisting of Algeria, Ecuador, Gabon, Indonesia, Iran, Iraq, Kuwait, Libya, Nigeria, Qatar, Saudi Arabia, United Arab Emirates and Venezuela) reduced its oil exports by 5% a month. The price of oil rocked from $3 to $12 a barrel. The countries of OPEC, dominated by the oil-rich Arab states, sought to use oil as a weapon in their fight against Israel. They hoped to force America to abandon its support of Israel, which had embarrassed the Arab nations by winning the Yom Kippur War in October of 1973.

Almost at once a global crisis arose. In the United States, long lines formed at gasoline stations. Weary motorists waited for hours to get enough gasoline for the next day's commute to work. The production schedules of industries all over the world had to be curtailed. Year-round daylight savings time was introduced in an attempt to cut back on electricity usage. President Richard Nixon asked Americans to lower the temperatures in their homes and offices by six degrees for a daytime average of 68 degrees. A 50-mile-per-hour speed limit was proposed, and a 55-mile-per-hour limit was adopted by Congress. In other words, Americans had to forfeit some of their customary comfort and carefree living.

In 1859, Colonel Edwin Drake struck oil near Titusville, Pennsylvania. Liquid gold, as petroleum came to be called, was originally used for lubrication and lighting. Later it became the fuel for America's drive-away, drive-in lifestyle. Today, the United States consumes 10,000 gallons of petroleum per second. "Oil was the answer to a peculiarly American prayer," wrote Richard J. Barnet in *The Lean Years: Politics in the Age of Scarcity*, "for it made possible the world's first society in perpetual motion. To achieve greatness, the United States spent oil as the empires of the past spent blood."[4]

The United States ranks third, after the Soviet Union and Saudia Arabia, among the world's oil producers. Because of the substantial quantity of petroleum available from American sources, the oil-rich countries of the Middle East could not at first dictate the world price of oil. So great was our use of oil, however, that by 1948 the United States was importing more oil than it was exporting. The Arab oil embargo of 1973 demonstrates not only the importance of oil to the world's economy, but the extent of America's dependence on foreign oil and our inability to guarantee a steady flow of oil at reasonable prices. America's reliance on imported oil had grown unnoticed by most Americans, and the oil embargo hit the nation with tremendous impact.

In the 1880s, the U.S. Geological Survey concluded that there was little or no chance that oil would be found in California, Kansas or Texas. Time proved the survey wrong, of course. Similarly, each time dire shortages of oil have been predicted, some new oil field has been found, allowing Americans to continue their wasteful use of oil resources.

Today, geologists estimate that the world's proven reserves of oil will last 30 years at current rates of consumption. The United States has eight years of proven reserves. A "proven reserve" is oil that is known to be available and recoverable using existing technology. Current opinion holds that there are few undiscovered large oil fields, but there is a continuing search for offshore oil off the Gulf Coast of Texas, in the North Sea, on the Georges Bank in the Atlantic, and other places.

Yet these are desperate measures. The unvarnished truth is that we are moving rapidly into a post-petroleum world. While there is general agreement that people must develop technologies to use renewable energy sources, there is little agreement as to which alternative is best. As we have worked to move from petroleum to other sources of energy, we have increasingly relied on natural gas as a resource.

Natural Gas

In 1825, the Marquis de Lafayette thought it something of a novelty when, in Fredonia, New York, he was invited to savor a meal cooked entirely by gas. Although widely regarded as a dangerous nuisance elsewhere, natural gas was used to light the street lamps of Fredonia. It was here that natural gas was first extracted commercially in the United States (in 1821) when local investors tapped into a pocket of natural gas and proceeded to sell it to the town and to a local inn.

Natural gas, a mixture of hydrocarbon gases (of which methane is the most important), is found in isolated deposits and near oil fields. When found close to oil, the flammable and explosive gas poses a danger to oil drilling. At first, natural gas was "flared" or burned off so workers could reach the petroleum. But when its uses and commercial value became better known, the gas was drawn off separately in pipelines. Now, because natural gas is the cleanest-burning fossil fuel, environmentalists are urging that it be thought of as a fuel of transition while we move into a post-petroleum era.

Methane, the major ingredient of natural gas, proved to have so many uses that it helped launch the petrochemical industry. Some brand-name synthetic fabrics, such as Orlon, Dacron and Nylon, are derived from natural gas, as is synthetic rubber and certain anesthetics, detergents and fertilizers. Natural gas is used in heating, cooling and cooking. Richard J. Barnet wrote that the use of natural gas changed "the face of suburbia.... Gone were the early morning furnace stokings, fumes, ashcans dragged down icy walkways, and grime, which were the hallmarks of coal civilization."[5]

Methane gas can be derived from cow manure and other sources, such as sugarcane residues and timber. And because these are renewable sources of energy, they may someday help expand the nation's energy store and contribute to the conservation of oil. Another technique for slowing the depletion of fossil fuel reserves is to dilute fuel oil with alcohol. This idea has been around for a long time, but the petroleum and gas companies have discouraged its use. The blended fuel has shortcomings—it is more expensive than plain gasoline and its use may require design changes in the engines of today's cars—but there is increasing interest in using alcohol derived from grain as a gasoline supplement. Brazil now distills a substantial amount of alcohol derived from the waste products of sugarcane production and mixes it with gasoline to stretch its fuel supply. Perhaps for certain applications, the blending of alcohol and gasoline is an idea whose time has come.

Promises and Perils of Nuclear Energy

On August 6, 1945, an American B-29 aircraft (the Enola Gay), flew 31,600 feet over the city of Hiroshima, Japan. At about 8:15 A.M., in a blinding flash that some described as brighter than the sun, an atomic bomb was dropped on the city. Over 60,000 people were killed. On August 9, a second atomic bomb was detonated over Nagasaki and 36,000 more were killed.

This use of nuclear energy demonstrates the promises and perils inherent in the commitment to this energy source. In an atomic bomb, an atom is split in a process called fission. The reaction releases nuclear energy in an uncontrolled explosion. In a nuclear reactor, by contrast, an atom is split in a controlled reaction that generates heat, which then is converted into power.

In 1951, the United States detonated a hydrogen bomb vastly more powerful than the atomic bomb dropped on Hiroshima. A hydrogen-bomb explosion is produced by fusing, or combining, two hydrogen-atom nuclei to form a single atom. In this process, an enormous amount of energy is released. We have not yet learned how to control the fusion process, although scientists have made some progress. To master this technology, we must develop here on Earth the same power that makes the stars shine. For now, though, when we speak of atomic energy, we refer to the process of fission.

Nuclear energy has many advantages over fossil-fuel energy. It is cheaper than coal or oil, and nuclear plants do not release chemical pollutants such as dust or sulfur dioxide into the air. Thus, nuclear energy does not contribute to acid rain or to the greenhouse effect. Nuclear power can help free the nation from its dependence on unreliable Middle Eastern supplies of oil. A new generation of breeder reactors may assure ample energy for hundreds of years and even allow for substantial increases in the demands for electricity. With all these advantages, some believe we should take a closer look at nuclear energy.

Nuclear power in the United States is in deep trouble politically, however. Its base of public support has eroded because of genuine and widespread concern about the safety of nuclear-power plants. Alarm about radioactive waste and fallout has simmered below the surface ever since we learned how to split the atom.

On March 28, 1979, a nuclear reactor at Three Mile Island in Middletown, Pennsylvania, experienced a near disaster with its radioactive core. The Nuclear Regulatory Commission discovered that the plant came to within an hour of a catastrophic meltdown, which would have released enormous quantities of deadly radiation into the air. Some radiation did escape, and large numbers of people were evacuated. It was a close call, but nobody died directly as a result of the accident.

In 1986, the core of a nuclear reactor at Chernobyl, near the city of Kiev in the Soviet Union, melted down. Chernobyl was "...an excellent example of a true nuclear disaster: events were for a time completely beyond control; a gas explosion occurred, and the reactor core melted down; the release of radioactivity was very large," wrote Tom Johnson in *The New Republic*."[6] About 100 people died almost immediately from the explosion and large-scale exposure to radiation. Many experts were astonished that there weren't more deaths.

Just as troubling as these dramatic examples is the question of what to do with the highly radioactive waste produced by nuclear power plants. These wastes remain radioactive for thousands of years, and can cause cancer, birth defects and other health problems. Spent fuel can be reprocessed, reducing the amount of waste that has to be disposed of, but this yields plutonium, a dangerous ingredient in the manufacture of atomic bombs. How can the nation prevent plutonium from falling into the hands of irresponsible groups intent on creating global chaos?

There is a good deal of argument as to whether or not there is such a thing as a "safe" level of radioactivity. If there is no such level, then nuclear energy is not a real energy option. There does, however, seem to be hope among some scientists that nuclear power plants safe enough to protect the public can be built. However, the problem of storing radioactive waste remains.

The nuclear disasters at Three Mile Island and Chernobyl struck a sensitive nerve in the American people. Americans have also been distressed by the ineptitude of government and industry in the area of radioactive-waste disposal. Lasting concern has spurred the growth of an active antinuclear lobby that believes there is no room for compromise when it comes to nuclear energy. Because of the efforts of these groups, the construction of nuclear plants in the United States has come to a virtual standstill. Licensing new plants is tedious and slow, while building them has become so expensive that they are no longer cost-effective. As a result, the nation faces an uncertain energy future. Other nations, especially those with limited supplies of fossil fuel (such as France and Japan), have pushed to develop substantial nuclear-energy facilities. The Soviet Union will probably move rapidly to develop its own nuclear-energy capacity as well, and will likely surpass the United States as a nuclear force. Nuclear energy from fusion is many decades away. In the meantime, the search goes on for energy alternatives.

Alternative Energy Sources

When nuclear energy was in its infancy, many were sure the worry about available energy resources was over. Some even predicted that energy would be so cheap that it could be distributed free of charge. Visions of a worry-free, cost-free energy future, however, did not materialize. Nuclear energy, we have seen, is fraught with difficulties, and so we have turned once again to age-old energy sources.

Today, many look to the renewable resources of wind, water and solar (sun) power to meet the world's future energy needs. Strictly speaking, these forces are not energy "sources." Instead, they are simply the steady flow of the sun's rays, rivers' currents, ocean tides, and blowing winds. Because they replenish themselves naturally, they have long been seen as a means by which modern society can provide for its energy needs.

It would be a mistake, however, to assume that using these forces of renewable energy would require a return to a simpler life. If these non-polluting forms of energy are to be harnessed in the service of modern life, communities will need to take a highly sophisticated, high-tech approach while working within the laws and limits of nature.

Wind Power

Wind power is probably the oldest form of energy used by people. Wind moved ships at sea and performed such mechanical tasks as pumping water and grinding grain into feed and flour. In early America, the windmill was an important means of industrial power, one that remained important through the first quarter of the 19th century. The oil crises of the 1970s triggered a renewed interest in wind as a source of energy. Today, wind power is the fastest-growing alternative means of generating electricity after nuclear energy. But our interest in it waxes and wanes with the price of oil. When the price of oil is low, and wind power no longer seems cost-effective, interest drops. When, however, oil prices rise or the Middle East becomes particularly unstable, development of wind power proceeds.

However, if you've ever tried sailing a boat, you have a pretty good idea of how elusive wind power can be. You know how difficult it is to find enough wind, where you want it, to steer your vessel in the direction you wish to go. On a much larger scale, the erratic characteristics of wind power can be partially overcome. The wind's vagaries would suggest, however, that while wind can be an important contributor to our energy needs, it is likely to remain an auxiliary source of power for many decades to come. Wind power also remains more costly than energy from fossil fuels, hydroelectric sources or nuclear fission. Nevertheless, according to the United States Department of Energy, by 1995 wind power may generate 10 billion kilowatt-hours, and 70 billion by 2005, in the United States.

Water Power

In early America, rushing water from streams and rivers turned waterwheels that powered machinery in textile factories. During the 1930s, huge dams were built in the western United States to help irrigate arid lands, generate electric power, control floods, improve river navigation, and safeguard the environment. The nation's reliance on fossil fuels increased, however, and interest in hydroelectric power faded. Nevertheless, hydroelectric power has so many desirable characteristics that it always comes up in discussions of America's energy needs. Hydropower does not use up water and does not contribute to air pollution, so it remains a highly desirable energy alternative.

In more recent years, a renewed interest in hydropower has developed, leading to the construction of small dams that generate moderate amounts of electricity. These dams offer many advantages, including environmental and aesthetic appeal. Construction of small, privately owned dams was encouraged by the Public Utilities Regulatory Policies Act of 1978, which requires public and private utilities to buy electric power from small producers. Many new dams are now being considered.

Solar Power

It is little wonder that many of our ancestors worshipped the sun. They had an instinctive sense of its power and viewed it as the source of life. Solar power is both the oldest and the newest energy alternative. It is the oldest in the sense that the fossil fuels we use today were first created by the action of the sun's rays on dead plants millions of years ago. Solar energy is also a component of wind power. The heat of the sun causes the wind to blow in one direction and then in another. Similarly, the sun plays a role in water power, causing evaporation and influencing the ebb and flow of the tides. Solar power, in the sense of the direct utilization of the sun's rays for generating electricity, is favored by many environmentalists.

Immediately after the oil crisis of 1973, interest in solar energy soared, and people started to look to the sun for their energy needs. Industry and government both encouraged further research. They saw a growing market for solar technology. But as the price of oil declined at the end of the oil crisis, interest in solar power ebbed.

In 1979, President Jimmy Carter installed a solar water heater in the White House as a symbolic expression of our determination to free ourselves of dependence on foreign oil. The heater supplied the kitchen with hot water for seven years. But in 1986, during the Reagan administration, laborers dismantled the solar collector to repair the roof. It was never replaced—a fact that is symbolic of the nation's shortsightedness on energy matters.

Solar energy is inexhaustible and non-polluting, it produces no radioactive waste, and it cannot be disrupted by global bickering. The technology to harness it is available, and ranges from the simple to the highly sophisticated. Using solar energy "passively" involves the simple absorption and storage of the sun's warmth in a building's structure. Passive solar buildings generally feature south-facing windows or insulated glass panels that admit sunlight to the building, where it warms the walls and floors.

In an active solar system, the sun's rays are collected by panels fitted with liquid- or air-filled tubes. The heat is usually transferred to a storage tank, from which it is then pumped when needed.

A photovoltaic or solar cell converts the energy of the sun into electricity. Photovoltaic cells in space can generate ten times the electricity similar cells can generate on Earth. But the cost of manufacturing these cells is very high. There is considerable doubt that in the near future the cost can be reduced enough to make electricity from the sun cost-competitive with that produced from coal or oil.

For a time, the government offered tax credits for people investing in solar-heating systems. However, such incentives to convert to solar energy have been drastically reduced. Will it take war, or another oil embargo, to spur Americans to develop and utilize environmentally safe, renewable sources of energy?

Geothermal energy from the core of the earth and biomass energy derived from plant and animal wastes offer other important non-polluting energy alternatives. But, although alternatives to both nuclear energy and fossil fuels appear plentiful, there has been a good deal of foot-dragging on the part of the government and industry in their development. As Richard J. Barnet wrote, the question before the nation is,"Who shall collect sunbeams? Who shall gather the wind?"[8]

Notes

Citations in the notes are brief. Full citations appear in the bibliography.

CHAPTER FIVE

1. Hedley, 5.
2. Cruver, 29.
3. Halberstam, 14
4. Barnet, 22.
5. Barnet, 68.
6. Johnson, 16.
7. D'Allesandro, 34.
8. Barnet, 109.

CHAPTER SIX

Not in My Backyard: *The Political Problems of Environmentalism*

In 1987, the community of Islip, New York, decided that in order to save space for residential garbage, its dumps would no longer accept commercial trash. This simple and seemingly logical decision triggered a crisis that Islip's well- intentioned community leaders did not anticipate.

To get rid of their trash, local business owners paid a waste disposal company to haul it way. The company chartered the tug *Break of Dawn* to push the barge *Mobro 4000*, loaded with 3,186 tons of commercial trash, to a destination in North Carolina. But the *Mobro 4000* spent 187 days at sea, rejected by community after community. Finally, the trash was incinerated in Brooklyn, New York, and the ashes were sent back to Islip.[1]

America throws away more trash per capita than any other nation in the world. For example:

- Each and every day of 1960, every man, woman and child in the country discarded 2.9 pounds of trash. By 1988, the amount of trash generated daily by each American had grown to 3.5 pounds, and by the year 2000 it is projected to reach 6.5 pounds.

- New Yorkers annually produce enough garbage to cover all 850 acres of Central Park to a depth of 13 feet.
- Each year, Americans throw away 16 billion disposable diapers, 1.6 billion pens and 2 billion razors. Included are 1,300 pounds of household waste per person.
- The 160 million tons a year of solid waste that Americans generate could fill a line of 10-ton garbage trucks extending halfway to the moon.[2]

Do these statistics mean that there is a garbage crisis in America? The 3.5 pounds of garbage currently generated daily by each person in the United States is not the highest it has been. At the turn of the century in New York, residents disposed of about 4.1 pounds of trash each day. A significant percentage of this figure consisted of ash produced by the burning of coal. Nevertheless, the problem of getting rid of garbage is a serious and growing one.

American industry encourages wastefulness by convincing consumers that they need all the "latest" or "newest" products. This selling technique is used for products as varied as deodorant and automobiles. In both cases, the latest and newest represent marginal improvements over last year's models. Nonetheless, consumers take the bait and prematurely throw away goods that are still useful. Thus, graveyards of automobiles and landfills packed with all kinds of usable items dot the American landscape.

In his futuristic novel *Brave New World*, published in 1932, Aldous Huxley envisioned a society in which each citizen was forced to consume a certain quota of goods each year. In many ways Huxley's forecast has been borne out. The throwaway character of modern America was evident as early as the 1950s.

Growing numbers of families are headed either by single working parents or couples in which both are breadwinners. These busy people seek convenience, usefulness and time-saving features in the products they buy. "Waste not, want not" may at one time have been an American value, but for decades now America has been a disposable society. Frozen dinners that can be popped into a microwave and eaten within a few minutes may be a boon to busy people, but the throwaway container in which it is packaged adds to the mountains of trash we produce.

Environmentalists use the term "garbage" to refer to organic waste. It consists of uneaten food, the inedible portions of the food items (such as banana peels, orange rinds or peach pits), fat trimmed from meat, and the bones of fowl or meat. "Refuse" is an all-embracing term that includes garbage as well as ashes, plastic, paper, rags, grass cuttings, leaves, wood, dead animals, metal cans, glass bottles, abandoned cars and tires, and worn-out appliances. "Solid waste" is the term environmentalists apply to everything we choose to throw out.

The number of items we use only once and then throw away is enormous. Some of the junk we throw out seems never to go away. One very good example is the disposable diaper. Chances are that the disposable diapers that you wore as an infant are still around—someplace. Disposable diapers pose two problems to the environment: the human waste discarded with the diaper is a hazard to the health of the community. Moreover, these diapers are made from materials that never decompose by natural biological means. Such products are called non-biodegradable.

Many of the common products you use every day, such as the package in which your fast-food hamburger is delivered, the plastic bag you get in the supermarket, the styrofoam cup in which your take-out coffee may be served, and the disposable plates you take on a picnic do not decompose even many decades after being thrown into a garbage dump. These synthetic and plastic products represent less than 10% of the waste we discard, but they complicate the way we collect and dispose of garbage.

Non-biodegradable products constitute a danger for reasons other than their direct contribution to environmental pollution. They are made using raw materials that in some cases include the carcinogen benzene. They also endanger the health of the community, because the blowing agents used in making some of them—styrofoam cups, for example—consist of chlorofluorocarbons, a major cause of ozone depletion. The chemical industry is taking steps to discover alternative ways of making these highly useful and convenient products less environmentally hazardous. However, many environmentalists feel that the only way to solve the problem of non-biodegradable packaging is to stop using it altogether.

If you ever take out your family's garbage, you know what a troublesome task this can be. The garbage container may leak. It may contain metal, glass, newspaper, packaging of various kinds, and leftover food. It all gets mixed together in your garbage can and is eventually loaded into a garbage truck and taken away. But where does it go? There are three ways to treat the garbage you throw out. It may be dumped in a landfill, it may be recycled, or it may be incinerated (burned). Each of these techniques has important consequences for the environment.

Landfills

"The largest landfill in the world sprawls over 3,000 acres at Fresh Kills, on Staten Island. It is an awesome sight:...a surreal procession of garbage hills and garbage dales and glittering garbage plains where bulldozers scuttle about like oversized scavenger beetles, towing trolleys full of trash," wrote J. Tevere MacFayden in *Atlantic* magazine.[3] About 80% of the nation's solid waste is dumped in such landfills, but these are filling up rapidly. The search is on for either new facilities or for new means of disposing of the ordinary rubbish discarded each day.

A landfill is nothing more than a sophisticated garbage dump. At one time, everyday trash was simply dumped in a designated area. In a landfill, the waste so deposited is covered by earth movers that landscape the dump and encourage biodegradable materials to decompose. Landfills, however, are rapidly approaching their saturation point. The number of landfills in the United States has fallen 30% since 1984, and their number will continue to decline as they fill up and are closed. Since 1978, about half of the landfills in the country have been filled and closed. In the next five to ten years, one-third of the remaining landfills will be filled up and forced to close.

Finding a place for new landfills is difficult, because no community wants a landfill in its backyard. Environmentally conscious Americans worry about the possible hazards from leaking acids, poisonous fumes, or contamination of underground water that can accompany a landfill. Some believe that a landfill has the tendency to depress nearby home and land values and attract "undesirable" people.

In 1988, the Environmental Protection Agency imposed regulations on landfills, requiring them to monitor hazardous wastes and methane gas. Landfills are also required to control rats, insects, fire and odors. These regulations are not unreasonable, but they will hasten the closing of landfills and enormously increase the cost of dumping garbage in them. Various regulations have already increased dumping fees tenfold, and further increases are inevitable, once the regulations are fully implemented in 1991. By 1992, the EPA wants to reduce the amount of garbage shipped to landfills by 25% and to see more trash recycled.

Recycling

Because landfills by themselves cannot handle the growing tide of trash, some urge that much more garbage be recycled. Recycling is an old idea. In the 1930s, Los Angeles recycled 30% of its solid waste. During World War II, aluminum pots and pans were collected to be recycled for the war effort, and 43% of newsprint was recycled. When the crisis ended, Americans found it easier to go back to throwing out what they could not use.[4]

For recycling to be effective, American families must follow the example of the Japanese, who carefully separate their garbage into batches of paper, glass, metal and so on. Unless careful separation takes place, recycling is unlikely to be successful. It is very difficult, however, to convince people to sift and sort their garbage. It is also very difficult to enforce sorting laws, and impose penalties. Local authorities cannot realistically refuse to pick up garbage from uncooperative families. Sorting, however, does nothing to solve the problem of trash that cannot readily be recycled. Disposable diapers, coated papers, plastic items, and discarded refrigerators or air conditioners that contain hazardous wastes cannot be recycled.

In addition, markets for recycled materials need to be found. Many manufacturers believe that working with new materials is more economical and reliable and allows them to make better-quality products. "Numerous studies have shown that even if all 'realistic recyclables' were recycled, more than 70% of the trash would remain.... Mandatory [recycling] approaches may merely yield neat bales of landfills, recycling the problem rather than the waste," according to Michael H. Levin, a reporter for the *Wall Street Journal.*

Recycling is one of those ideas that should work—and which does work in some societies. But can it work in America? Americans would need to change their wasteful lifestyles. Many people would no doubt agree to separate their garbage into categories, but that's not the problem. The problem is that we are addicted to packaged goods of all kinds. We do not buy unpackaged loaves of bread, as do the French; we do not bring our own shopping bags when we go to the market; and we tend to shop at supermarkets, where most of the merchandise is prepackaged, instead of at farmers' markets, where the merchandise is sold unpackaged.

Manufacturers fight laws that require consumers to return cans and bottles for recycling. The government offers tax breaks for using new materials, thereby encouraging greater consumption. There are 15% depletion allowances for using iron ore rather than recycled iron. Allowances of 22% for oil, aluminum and natural gas also exist. Harvested forests are capital gain for timber firms. You can see why widespread recycling is not likely to happen soon.[6]

Nonetheless, environmentalist Barry Commoner continues to urge the nation to make recycling a major element in its program of solid-waste disposal. On the basis of an experiment involving 100 families in East Hampton, Long Island, he is convinced that most of the garbage in New York City could be recycled. In April, 1989, New York City adopted a recycling law. Since July 1, 1989, New York's households have had to separate their trash. Newspapers, cardboard and magazines have to be bundled. A separate container has to be provided for cans and glass bottles. The goal is to recycle 25% of New York's garbage by 1994.

Incineration

In the 1970s, incinerating garbage was a common practice for those living in apartment houses. They simply bagged their garbage and tossed it down a chute that opened into a furnace. The furnaces burned the waste, and sanitation workers hauled the ashes away. In order to reduce air pollution, this practice was prohibited, and the incineration of garbage has nearly ceased. However, because people now realize that they might soon choke on their own waste, there has been a renewed interest in community-wide incineration in place of some landfills.

However, incineration does not involve merely the burning of waste. Instead, in a process known as "resource recovery," incineration can convert trash into energy. The community of Westley, California, has the distinction of housing the world's largest pile of discarded tires, some 40 million of the estimated 2 billion in this country. These tires are being burned as fuel and transformed into electricity by Pacific Gas and Electric, the local utility. "One tire," says an executive involved in the program, "can serve the energy needs of the average northern California household for a day."[7] The incineration generates energy for the public and cash for the entrepreneurs, but it also spews noxious nitrous oxides into the air. The company has, however, installed sophisticated scrubbers and other smog control systems that appear to reduce these emissions to a minimum.

While "resource recovery" seems to be a light at the end of the garbage tunnel, the huge incineration plants necessary for the process are likely to be as unwelcome as landfills. "Not in My Backyard" is the continuing cry of Americans who fear poisonous emissions, noise and odors. In short, neither landfills nor "resource-recovery" facilities are thought of as attractive neighbors.

Sewage

To 19th-century Americans, cholera was a disease feared much as Acquired Immune Deficiency Syndrome (AIDS) is today. Men and women had not yet learned that cholera was caused by a lack of adequate sanitation and a failure to keep drinking water pure. Periodic outbreaks of cholera in 1832, 1849 and 1866 forced urban centers to address the problems of community health.

In 1842, New York City established a large supply of pure drinking water by building the Croton system of reservoirs and aqueducts. It was the first major American city to do so. It was not until the end of the 19th century that the standard of one private, sanitary toilet per family was established in U.S. cities. Today, gains in public sanitation are being eroded by our indifference to the handling of sewage.

Each American generates 60 gallons of wastewater daily. Meanwhile, the sewage systems in most major cities still rely on facilities installed almost 100 years ago. In heavy storms, existing city sewers in many cities are often unable to handle the runoff of rainfall in addition to their regular volume of sewage. For example, Boston Harbor is annually the recipient of five billion gallons of raw sewage and storm runoff.[8]

Boston Harbor is one of many into which scuba divers today fear to dive. A 1988 article in *Newsweek* reported, "Scuba divers talk of swimming through clouds of toilet paper and half-dissolved feces, of bay bottoms covered by a foul and toxic combination of sediment, sewage and petrochemical waste appropriately known as 'black mayonnaise.'[9] America's offhand approach to sewage treatment has fouled Chesapeake Bay, the Gulf of Mexico, San Francisco Bay, Puget Sound and many other beautiful harbors. Using their water or swimming in them now are hazardous to human health.

In New York Harbor, partially refined sewage known as sludge is dumped at the foot of the Statue of Liberty. If the Lady one day loosens her grip on her torch, it will be to hold her nose. New York City began dumping in 1938. Today, along with eight communities on Long Island and New Jersey, the city accounts for 95% of the country's municipal-sewage ocean-dumping.

In 1981, the Environmental Protection Agency officially prohibited sludge-dumping. But in the case of New York City, the EPA could not make the rule stick. The city, along with other dumping communities, won the fight, although they agreed to dump farther out to sea. Today, sludge boats head out to sea loaded with 710,000 gallons each of treated human waste. There, "bobbing in gentle ocean swells," according to *Newsweek*, "the boats slowly zigzag across a patch of ocean 106 miles offshore, releasing their cargo in jet-black plumes."[10]

The days of ocean dumping may soon be over, as sentiment mounts for it to end. But, as in the case of landfills and incinerators, there are few alternative methods of sewage disposal. New and more sophisticated sewage-treatment plants are necessary.

Sewage-treatment plants fall into three categories. Primary plants remove about 30% of the pollution, secondary treatment makes the waste about 85% clean, and advanced-treatment plants can remove about 95% of the pollutants from sewage. However, sewage-treatment plants are very costly. Communities are looking for other, less expensive and simpler alternatives.

Among the alternatives is low-cost sewage treatment, with which a number of smaller communities are now experimenting. These treatments involve creating artificial wetlands that filter pollutants much as natural wetlands do. Although relatively inexpensive, such artificial wetlands can serve only small communities and require a great deal of land. Farmers, however, are looking to artificial wetlands as a means of cleansing pesticide residues from the runoff of their irrigation processes.

The treatment of sewage and solid waste is a political hot potato. Responsibility is passed around as each community seeks desperately to put the financial burden on the back of its neighbor. All would agree that cleaning up after ourselves is a burden that must be shared, but apportioning the work equitably is very difficult indeed.

Hazardous Wastes: Ticking Time Bombs

Many common household chemicals—cleaners, paints, solvents—are hazardous to your health. They become particularly hazardous if they are disposed of carelessly. For example, were you to pour paint remover down the drain, dump silver polish into the ground, or throw a can of ant killer into your garbage can, these chemicals would eventually find their way into the water you drink, albeit in very small amounts. Many everyday household products simply do not break down when they are flushed into the toilet, nor do they become harmless at a water-treatment plant. Instead, they gradually make their way into the groundwater. Even minute quantities of such pollutants can accumulate over a period of time and contribute to cancer, birth defects and other illnesses.

Most hazardous wastes are by-products of the chemical production. They include dioxin, a by-product of the manufacture of plastics, dyes, pesticides and herbicides, and PCBs (polychlorinated biphenyls) used in electrical-transformer production. Because these and other chemical wastes are carcinogens, disposing of them so as to minimize health risks is complex and expensive. Because the harm caused by these and other hazardous wastes may not take effect for years, hazardous wastes have sometimes been described as "ticking time bombs."

In 1976, in belated recognition that hazardous-waste disposal was a serious problem, Congress passed the Resource Conservation and Recovery Act. It was the first federal law to deal with hazardous waste. The Environmental Protection Agency has set aside billions of dollars in a "superfund" to clean up abandoned hazardous-waste sites. Only a few sites, though, have actually been cleaned up.

Three hundred million tons of hazardous waste is produced in the United States annually. How to get rid of this waste safely remains a major environmental headache. As regulations governing the disposal of hazardous wastes become more complex, private entrepreneurs have entered the waste-removal business. Dumping hazardous waste in a landfill is no longer legal, so toxic sludge is often dumped into special lagoons lined with clay or plastic. Some hazardous waste is burned in incinerators, and the ash is buried. A more sophisticated approach is to mix the hazardous waste with other chemicals so as to make it non-toxic. Some hazardous wastes are also recycled.

The EPA has embarked on a program called "waste minimization." This program encourages industry to find alternative ways of manufacturing its products that use fewer hazardous materials. The program also offers incentives for recycling hazardous waste into other profitable products. One company has devised a way to recycle paraffin wax, produced during the manufacture of chewing gum, to make crayons or firelogs.[11]

Radioactive Waste

When nuclear energy was first harnessed, there was widespread optimism that a cheap and unlimited source of energy had at last been found. It was thought that the problem of radioactive-waste disposal was one that could be easily handled. As a result, the nation embarked on a nuclear energy program prematurely. Nuclear power plants were built without adequate provision for the storage or removal of spent fuels that emit a dangerous level of radiation. And we have yet to identify any effective way of dealing with radioactive waste.

An even bigger problem lies with the nation's nuclear defense program. Nuclear weapons facilities produce the components of nuclear missiles, bombs, warships and submarines. These plants, however, have also produced America's most dangerous garbage heap—enormous quantities of radioactive waste.

What to do with radioactive waste remains an essentially unsolved problem. A permanent underground site for the storage of spent fuels was agreed upon at Carlsbad, New Mexico. There, more than 2,000 feet below the New Mexican desert in salt beds many millions of years old, radioactive wastes were to be stored. However, reevaluation of the Waste Isolation Pilot

Program (WIPP), as the site is called, demonstrated that there may be too much water in the salt beds, and as a result radioactive contamination may eventually reach the surface. While the Carlsbad nuclear-waste storage facility will not be abandoned, it will be filled very slowly over the next five years, so that what happens to the nuclear wastes stored there can be monitored.

In the search for a new underground facility for radioactive nuclear waste, attention shifted to the Yucca Mountain site in Nevada in 1988. Here in the Nevada desert, about 110 miles west and slightly north of Las Vegas, on land owned by the federal government, a billion dollars will be spent to determine the feasibility of this location. But Nevada is not enthusiastic about the prospect of storing spent nuclear fuels in its backyard.

The problem of disposing of radioactive waste is a significant component of a larger question: What can Americans do about wastes from *all* sources? The wastes of a society nearing the 21st century cannot be managed with facilities designed for the 19th century. Technology, it appears, has created tremendous problems by generating new forms of trash. Yet it also holds out the promise of a better quality of life. And the solutions to these problems will no doubt require new technologies, and their thoughtful application by all of America's citizens.

Notes

Citations in the notes are brief. Full citations appear in the bibliography.

CHAPTER SIX

1. *New York Times*, October 23, 1988.
2. Levin.
3. MacFayden, 20.

4. Steinhart, 107.
5. Levin.
6. Steinhart.
7. Weber, 78.
8. Marx, 125.
9. *Newsweek,* "Don't Go Near the Water," August 1, 1988, 43.
10. *Newsweek,* "Don't Go Near the Water," August 1, 1988, 46.
11. Environmental Protection Agency, 20.

CHAPTER SEVEN

The Crown Jewels: *The National Parks and Wilderness Areas*

If America were a monarchy, its national parks and wilderness areas could be thought of as the nation's crown jewels. However, since America is a republic, the national parks and wildlife areas are perhaps better thought of as natural museums. The flora and fauna that are characteristic of an area are preserved in national parks as a reminder of what the American landscape was like in the early days of the republic.

"The history of America," declared President John Fitzgerald Kennedy in 1963, "has been the story of Americans seizing, using, squandering and, belatedly, protecting...their rich natural heritage." The low point of America's concern for its natural heritage occurred about one hundred years before President Kennedy made these remarks.

By the latter half of the 19th century, millions of acres of America's forests had been cut down. Farmers had cleared the land for agriculture, railroads and steamships had used the timber for fuel, and railroad expansion had called for vast quantities of wooden railroad ties on which to lay steel tracks. Rapidly growing cities had an insatiable demand for lumber. Fires, some deliberately and maliciously set, destroyed as much as 25 million acres of forest each year. Government corruption was widespread and unscrupulous, as federal bureaucrats and politicians allowed industry to ravage the country's natural resources.

Prophets of Preservation

There were prophets in the land who warned of the need to adopt land- and wildlife-preservation policies before it was too late. There were those who decried the wanton destruction, and whose cries were eventually heard by a reluctant Congress and president. One piece of legislation after another perpetuated the destruction of the nation's natural bounty, but as if to appease the government's conscience, other pieces of legislation sought to preserve that bounty in perpetuity. "The National Park System was invented as an utterly unique idea in American history, created in the dead center of an era of careless exploitation," wrote T. H. Watkins in *Wilderness* magazine.[1]

An artist was among the early prophets of preservation of our natural landscape—George Catlin, whose paintings of the American Indian are priceless today. In 1832, he visited what would become South Dakota, and envisioned a park there. But at the time, parks were carefully cultivated to conform to the human vision of beauty. A park in those days was supposed to be an oasis of serenity, a Garden of Eden, a miniature paradise. Catlin added a new dimension to the concept of the park: He urged that land be set aside in its natural state without being contoured to fit with the tastes of people. The park he had in mind was to be no human creation.

Catlin was not alone in recommending that the natural landscape of America be preserved for others to experience as he had. John James Audubon, the famous painter of American birds, lamented the disappearance of the "noble forest." Other important voices were those of the New England transcendentalists, who held that God lived in nature and that the truest religious expression was one in harmony with all natural things. Ralph Waldo Emerson and Henry David Thoreau are perhaps the best known of those who held this view.

George P. Marsh, Major J. W. Powell and John Muir were among other important naturalists urging that lands be set aside in their pristine state. These men dreamed of the way the American landscape could look if it was handled with care and concern. But they were not just dreamers; they were doers as well. They fought sophisticated political battles so that other Americans might see what they had seen of America's unaltered natural domain. As a result of their efforts, Congress established Yellowstone National Park in 1872.

It is something of an irony that the administration of Ulysses Grant, otherwise known for its widespread corruption and its policy of giving away public land to industry, should take the initiative to establish the world's first national park. After this precedent, other parks would follow.

In 1867, John Muir, the founder of the Sierra Club, began a 1,000-mile walk from Wisconsin to the Gulf of Mexico. He later studied the Yosemite Valley, which President Abraham Lincoln had ceded to the State of California on the condition that its giant sequoia trees be protected from commercial exploitation. Muir wrote about what he saw and published his articles in influential eastern magazines. By 1890, spurred by Muir's work, Congress created three more national parks: Yosemite National Park, General Grant National Park, and Sequoia National Park.

In 1869, General J. W. Powell, later to become the founder of the U.S. Geological Survey, made the first descent into the canyon of the Colorado River. The rich geological research he conducted helped convince the United States government to establish Grand

Canyon National Park in 1908. It was expanded in 1919. Grand Teton National Park was established in 1929, and with the help of industrialist John D. Rockefeller, Jr., the park area was enlarged. Grant Teton is part of the extraordinary and immense Yellowstone ecosystem.

The American Wilderness As a Fountain of Life

On September 3, 1964, President Lyndon Johnson signed into law the National Wilderness Preservation System. In doing so, he added a new dimension to the national park concept. In such wilderness areas, all forms of motorized transport, roads and buildings are forbidden. In some of these areas, limited cattle-grazing and mining are permitted. A national park seeks to make natural wonders accessible to visitors by providing roads, parking, accommodations, food services, lavatories and concession facilities; a wilderness area seeks to preserve the environment in as unspoiled a state as possible. Backpacking, horseback- riding and canoeing are among the activities in which visitors to a wilderness area may participate, but automobiles and other such amenities must be left behind.

The earliest American pioneers believed that the wilderness had aesthetic value that deserved to be retained. Daniel Boone, (1734–1820) for example, extolled the superiority of natural beauty over the manmade elegance of the city. The poet William Cullen Bryant (1794–1878) found in the American wilderness the source for much of his inspiration, particularly in his 1825 work, "A Forest Hymn." And American novelist James Fenimore Cooper, (1789–1851) author of *The Leatherstocking Tales*, also discussed the relationship of man and nature in his work.

But in 1831, when French social critic Alexis de Tocqueville (1805–1859) sought to explore the frontier wilderness for the sheer pleasure of it, his American guides could not believe there was any pleasure to be found in the wilderness. To these men, the wilderness was merely something to be cleared away.

Among those who helped us to see the wilderness in a different light was Aldo Leopold (1887–1948), forester, scholar, philosopher and ecologist. He proposed that, not only did human beings have rights, but the land and all the creatures who shared the land with humans had rights, too. Leopold, the patron saint of the Wilderness Society, was a member of the National Forest Service, which sought to protect and preserve large areas of land. The first such wilderness was the Gila National Park in New Mexico, which was set aside in 1924.

Conservation is the attempt to save the environment for future use. That which is saved offers some economic benefit or utility. The preservationist, on the other hand, seeks to save the environment for its own sake, regardless of the potential economic value of the flora and fauna being saved. Of what value is a wild bear, a wood-eating termite, a fruit fly, a boll weevil or a fire ant? Preservationists such as Leopold, however, insist that if we confine our interests to land and animals of economic value, we would save only a mere fraction of the natural world. In other words, to rely on economics as a basis for protecting the environment is to ignore the important non-economic role many plant and animal species play in nature.

Wildlands also have psychological and spiritual value. The Boy Scouts, the Girl Scouts, and organizations such as Outward Bound all help people to "build character" through facing challenges in the wild. And many large corporations have used wilderness experience as a technique of developing self-reliance and team spirit among their executives.

Noah's Unfinished Task: Wildlife Preservation

The mid-19th century was a dark time for the nation's environmental policy. Great tracts of unspoiled land were recklessly abused or given away to industry. Wildlife devastation was equally appalling at that time. In 1810, for instance, there were about five billion passenger pigeons in the United States. They made up the most abundant wildlife species on the continent. But the nation set out to destroy this species, and conducted savage "pigeon kills." In 1878, at one such kill, 300 tons of birds were exterminated in a month.

When the tsar of Russia agreed in 1867 to sell Alaska to the United States for $7.6 million, there were about five million seals living there. In the first twenty years of its operation, the Alaska Commercial Company, which received the government's franchise to hunt the seals, killed so many seals that the skins harvested alone could have repaid the entire cost of the Alaska Purchase.

The American bison, or buffalo, a symbol of the American West, shared the fate of the seals. The 50 to 100 million bison that once roamed the Great Plains were a source of food for the American Indians who lived there. The hides kept the Indians warm, the horns were used as drinking vessels, and the bones were honed to fashion weapons and tools. So dependent were the Indians on the bison that the U.S. Army recommended the destruction of the animals as a means of conquering the Indians.

Professional hunters such as William Frederick Cody (Buffalo Bill), as well as many amateurs, shot the bison from railroad carriages, nearly wiping out the herds. In 1889 Congress officially ended the carnage, outlawing the slaughter of bison. The few remaining wild bison, about 90 in number, were moved to the protected lands of Yellowstone National Park.

Ever since they learned to use weapons in the hunt, humans have been the most dangerous predators on Earth. Especially as their skills and cunning have improved, human beings have become the major threat to other living things. To free land for farming, people have exterminated wildlife so that they could plant their crops. The growth of the population, the march of technology, rapid urbanization, the destruction of forests, the draining of marshes, the construction of huge dams for hydroelectric power, and the diversion of mighty rivers and lakes have all made the Earth less hospitable to animal life, both on land and in water.

The United States government became concerned about what was happening to the country's wildlife, but such concern was slow to surface. In 1894, hunting was prohibited in Yellowstone National Park. This established a precedent for the prohibition of hunting in all national parks. In 1903, Theodore Roosevelt signed an executive order creating Pelican Island Reservation on Florida's east coast, to protect the pelican colony there. In 1929, Congress passed the Migratory Bird Conservation Act, which provided for the acquisition of land and water areas for the protection of migratory waterfowl. In 1937, Mexico and Canada joined the United States to protect birds that fly across national boundaries to breed. In 1940, the Fish and Wildlife Service was established and given responsibility for the welfare of fish and animals outside of the national parks. And by 1964, a system of wildlife refuges had been established.

In recognition that even more needed to be done to save fish and wildlife, Congress passed the Endangered Species Act in 1974. As a result of this measure, the American alligator, the red wolf, the bald eagle and other species long abused were finally protected. But Americans had acted too late for such extinct species as the dusky seaside sparrow and the Palos Verde blue butterfly. There are 1,035 species, such as the California condor and the black-footed ferret, on the endangered list of the Fish and Wildlife Service. Half of these are found only in the United States. The government has made plans to save 275 of these species.

Today, there is growing sensitivity to the obligation that humankind has to treat other living things with respect. There is concern not only about animals in the wild, but also with how domesticated and farm animals, particularly those grown for food, fur and scientific experimentation, are treated. In "factory farms," many of these animals are treated cruelly, confined to small pens and force-fed to fatten them for market.

According to the Bible, God assigned Noah the task of building an ark to save animal life from a flood that wiped out humanity. It is evident, however, that Noah's task remains unfinished.

The Paradox of the Parks

If you were the director of the National Park Service, how would you solve this problem? Off-road vehicles (ORVs), mounted on bloated tires and powered by gasoline engines, can travel at about 30 miles per hour on any kind of terrain. Snowmobiles and dune buggies are examples of such vehicles. Should they be permitted in the national parks? This is a simple example of the kinds of decisions those in charge of the nation's 341 park areas need to make.

If you believe that the purpose of the national parks is to preserve the environment, then clearly you would rule against the use of noisy vehicles that disturb wildlife and tear up the terrain. If, on the other hand, you believe that the purpose of the national parks is to provide people access to nature, then you might be inclined to try to accommodate the ORVs.

The paradox of the national parks is how to make them accessible without destroying them. How much should the natural setting be altered to allow for sleeping accommodations, food sales, picnicking, camping, parking, roads, concessions, lavatory facilities and the like? James Watt, one of the most controversial secretaries of interior, served from 1981 to 1983 during the Reagan administration. He claimed to support the national parks, and bemoaned their rundown condition. However, he showed little interest in expanding the park systems: Quite the contrary—he thought there were too many parks. He had little concern for how the flora and fauna of the national parks fared. He wanted to satisfy the growing number of people who want easy access to the parks rather than preservation of the wilderness and wildlife for posterity.

How much access to the national parks is appropriate? What kinds of activities should be encouraged in the parks? For example, should single-engine airplane or helicopter rides be permitted in Grand Canyon National Park? The view of the park from the air is memorable, magnificent and breathtaking. But there is an environmental cost in noise and air pollution, which is contrary to the original purpose for which the parks were established. The aircraft frighten wildlife and give off poisonous carbon and sulfuric fumes, contributing to environmental destruction.

Most of the important problems of the environment we have so far discussed affect the national parks. Acid rain, pollution of aquifers, and indiscriminate use of pesticides and herbicides adversely affect the very quality of the national parks. In order to minimize the erosion of the parks' environment, the perimeter of the parks must be kept free of polluting industries and mining activities. If there were a limit on the kinds of activity that can be conducted within several miles of the park, what impact would that have on the nation's economy? How can we protect the environment on the one hand and promote economically vital industries on the other? A new breed of conservationist hopes to accomplish both goals through the pragmatic use of parkland and the maintenance of the ecological balance within the parks.

In a certain sense, the American system of national parks is a victim of its own success. The parks were chosen for their environmental uniqueness, as examples of the fragile relationship between land, water and living things. For that reason, they have become popular vacation spots. Precisely how much use can any park sustain with ecological damage? Visitors enter a park with the best of intentions. But they no longer light their fires without matches, skin squirrels and roast them at the end of a stick, and make a pouch of the squirrel's skin. Instead, they sleep in Dacron-filled sleeping bags, house themselves in nylon tents, use graphite-fiber fishing rods and fiberglass cross-country skis, and eat chemically processed freeze-dried food. They use a host of plastic products, including insulated styrofoam drinking cups and plates, and plastic utensils, which they bring to the parks or buy there. These non-biodegradable products pose a direct threat to the ecological balance the park was established to preserve.

In the summer of 1988, a huge fire swept through Yellowstone National Park. In response, the professionals of the National Park Service followed what the press called a "let-it-burn" policy. Where neither life nor property were in danger, and when the fire had started naturally by lightning, for instance, rather than by careless campers, the fire was allowed to burn itself out. Fire is a part of the natural ecological process by which dead and dying foliage is cleared away to make way for new growth. Most of the public did not understand the policy, and neither did many in Congress. Tourists, and states that depend on tourist dollars for their prosperity, see a charred forest as one that will not attract visitors and the money they spend. Industry sees only the loss of timber. But as research biologist Don Despain explained in the *New York Times Magazine*, "From an ecological point of view, there is no damage, there is still a forest ecosystem out there, nothing this park was established to preserve is lost. Dead trees may be a loss to housebuilders, but not to woodpeckers."[2]

Most experts believe the fire of 1988 will eventually help renew the ecosystem of the nation's oldest national park. It should also renew attitudes toward our system of national parks. A good deal of additional research is needed about the complexities of the ecosystems the parks contain. Park access needs to be balanced by wilderness and wildlife preservation.

Cecil Andrus, former secretary of the interior, wrote in *Wilderness* magazine, "I refuse to believe that the American people enter their parks primarily in search of motel rooms and hamburger stands, souvenir shops and filling stations, snowmobile trails and tennis courts. They come, and they bring their children with them, to find something more important: a kinship with the natural world around them, the awe which comes when human beings are in the presence of majestic beauty created by the forests which have shaped this planet. That is the experience we have inherited. That is the experience, intact and enlarged, which we are morally bound to pass on to generations after us."[3]

Because the national parks reflect the condition of the American environment at large, they reflect the nation's environmental attitudes. National and state governments, American business and industry, and American schools and individuals need to cultivate habits of living and thinking that will allow the nation to protect its environment more effectively. Each has a role to play in protecting the environment. Let us see what that role might be.

Notes

Citations in the notes are brief. Full citations appear in the bibliography.

CHAPTER SEVEN

1. Watkins, 3.
2. Matthiessen, 128.
3. Andrus, 44.

CHAPTER EIGHT

Cleaning Up: *Who Should Solve Our Woes?*

Few would disagree that the American environment needs to be cleaned up. But who is best equipped to do it? Should the job be left to government? If so, should the responsibility be taken by the federal government, or should the task be left to the states? If federal, state and local governments are to share the responsibility for environmental cleanup, how much governmental regulation is necessary? What should the nature of such regulation be? Do American business and industry have a responsibility to improve the environment? If so, what responsibility do they have? Since pollution recognizes no borders, how can nations work together to prevent the destruction of the planet Earth?

The Role of Government

The United States Department of the Interior describes itself in a brochure as "The Department of Everything Else." Congress established the Department of the Interior on March 3, 1849. One would think that the establishment of a cabinet department devoted to internal matters would generate little opposition. After all, in the 1840s, the nation was seeking to fulfill its "Manifest Destiny" to expand from the Atlantic to the Pacific. In 1845, Texas was annexed, and a border dispute with Britain over Oregon was resolved. War with Mexico ended in 1848, and the victorious United States annexed a large tract of Mexican land. Clearly, a department devoted to the dramatic concerns of an expanding nation would seem appropriate.

But the 1840s were years in which American domestic affairs centered around the problem of slavery. Some sought the end of slavery, while others sought its expansion into the newly acquired lands. Many feared that the Department of the Interior would usurp the powers of the states. Thus, the new department, set up to manage the nation's domestic affairs, became instead "the department of everything else," the "Great Miscellany," a "hydra-headed monster," and "a slop bucket for executive fragments."[1] It was not until the early 20th century that the department became concerned with the conservation of America's natural resources. The department assumed responsibility for parks, mines, land reclamation, fish and wildlife. The Department of the Interior does not have the power to establish a broad-based environmental effort, however. The Bureau of Land Management is under its authority, but the U.S. Forest Service is part of the Department of Agriculture.

Among the more environmentally conscious secretaries of the interior was Stewart Udall, who served during the Kennedy and Johnson administrations. He stated his concern in a book entitled *The Quiet Crisis*. "America today stands poised on a pinnacle of wealth and power," he wrote, "yet we live in a land of vanishing beauty, of increasing ugliness, of shrinking open space, and of an overall environment that is diminished daily by pollution, noise and blight."[2]

Udall sponsored federal environmental legislation such as the Clean Air Act of 1963, the Wilderness Act of 1964, the Land and Water Conservation Fund Act of 1965, the National Historic Preservation Act of 1966, the Wild and Scenic Rivers Act of 1968, and an amendment strengthening the federal Water Pollution Control Act of 1956. While he was secretary, the department established four national seashores along the Atlantic coast, undertook major pollution-abatement efforts on Lake Erie and the Hudson, Delaware and Potomac rivers, and initiated the national-capital beautification program sponsored by Lady Bird Johnson, wife of President Lyndon Johnson.

Under President Richard Nixon, the United States Environmental Protection Agency (EPA) was established. Unlike the departments of the interior or agriculture, the EPA is not a cabinet department but an independent agency. In principle at least, its sole client is the environment.

The EPA is responsible for assuring the nation of a supply of safe drinking water and ending the pollution of oceans, rivers, streams, lakes, estuaries, underground aquifers and wetlands. It has the responsibility of cleaning up the air and regulating the treatment of solid and hazardous waste. A "superfund" allocates money to the EPA so it can respond immediately to conditions that endanger human health or the environment. The agency has focused on the disposal of hazardous materials that have been improperly handled.

The Federal Government as Mr. Clean

The role of the federal government in protecting the environment has been endlessly debated, and no doubt will be debated as long as there are environmental problems. The problems faced by the EPA point up the difficulty of finding a place for the government in environmental affairs.

The EPA, on the one hand, must be sure that the enactment of its environmental regulations is justified, and that the results achieved will be worth the costs. Often, research is needed to ensure that a proposed remedy will not trigger other environmental problems. On the other hand, the EPA sometimes takes so long to act on a problem that it is often too late to do anything about it. Indeed, public pressure is usually what spurs the agency to action.

A comprehensive approach to the problems of the American environment has never been developed. Instead, crisis management has been the government's usual technique. At the federal level, environmental problems are primarily the responsibility of the EPA, but the Department of Agriculture and the Department of the Interior still have important roles to play. This disorganized bureaucracy has already addressed the "easy" problems, but the more complex ones remain. These include pesticide control; the disposal of solid, hazardous and radioactive waste; acid rain; and the greenhouse effect. The EPA, the Occupational Safety and Health Administration, and the Consumer Product Safety Commission all have responsibility for regulating substances believed to pose health risks. By the time such substances are identified, it is difficult to turn the situation around quickly. They must not only determine which pollutants cause cancer, birth defects or skin problems, they must also assess the precise degree of risk each poses.

If all hazardous substances were banned regardless of their level of risk, our standards of living would fall sharply. Economic dislocations could be substantial, and there could be a rise in unemployment. As a result, these agencies must weigh each decision carefully. The government's approach is, therefore, often far more cautious than environmentalists would like since they insist that the government is not doing enough. The government's point of view tends to be that there are so many constituencies to be served, so much data to be evaluated, so little money, and so much to be done, that it is doing all it can under the circumstances.

Pollution-control-and-abatement spending rose from $10 billion in 1970 to $75 billion in 1985. This represents an increase from 1% to 2% of the Gross National Product (GNP)—the value of all goods and services produced in the United States. Because of new regulations, the costs of pollution control could soon grow to 3% of the GNP. Are we spending enough? Probably not. But we can perhaps never spend enough on the environment. Environmental standards set a decade ago will most likely be insufficient for the decades ahead. Growing populations and improved living standards, however desirable these may be, will continue to generate additional pollution problems. The price of more people and growing affluence is more pollution.

One example of growing costs is the EPA's "superfund," a multibillion-dollar trust fund allocated by Congress in 1980. The EPA is authorized to use the fund to identify the country's worst toxic-waste sites and to force the responsible parties to clean them up. If the responsible parties cannot be found or refuse to undertake the cleanup, the EPA does the job and sues to recover the money. Only about three dozen sites have been cleaned up so far—out of 27,000 identified.[3]

This dismal situation has many causes. For one thing, no amount of money can erase the scars toxic waste has left on our environment. In addition, it is difficult to identify the appropriate cleanup technology for each case, whether incineration, chemical treatment or some other technique. The technology of cleanup is both costly and rapidly changing. As a result of these and other problems, the superfund has been held up as an example of a giant boondoggle. It has generated a great deal of activity, involving vast expenditure, but has yielded few results. While mismanagement and incompetence have undoubtedly contributed to the superfund's failure, the very character of our environmental management is also responsible. We must set meaningful environmental standards and find the will, the money and the technology to meet those standards.

The States as Mr. Clean

Much of the responsibility for the environment has been turned over to the state and local authorities. Many states are glad to assume the role of Mr. Clean, because they feel that they have a better understanding of the environmental needs of their region. They complain of what they see as inadequate financial support from Washington, but some of the most important innovations in pollution control are nonetheless occurring at the state level:

- California's Proposition 65 prohibits businesses employing ten or more people from discharging certain chemicals into the water or on land, where they might contaminate drinking water sources, unless they can show that their actions will pose no significant risk to the water supply.
- In New Jersey, the state legislature passed the Source Separation and Recycling Act, which aims to reduce demand for new landfills by recycling at least 25% of New Jersey's solid waste. Oregon, Connecticut and Rhode Island have similar laws, but New Jersey's goes the furthest.
- In May 1987 Iowa enacted comprehensive legislation to protect its groundwater from synthetic chemicals, fertilizers, pesticides and industrial waste. The law offers incentives for industry and communities to research alternative waste-disposal methods.
- The states of New York, Connecticut, New Jersey, Maine, Massachusetts, New Hampshire, Rhode Island and Vermont agreed to restrict the amount of gasoline vapor that can be emitted by automobiles.
- Virginia, Maryland, Pennsylvania and the District of Columbia are cooperating in efforts to clean up the badly polluted Chesapeake Bay.

These initiatives are good examples of what can be done by states to improve the quality of their environment. But environmental problems do not stop at state lines. A comprehensive approach to the nation's environmental problems can hardly be expected from 50 separate state governments.

The Greening of the Invisible Hand

In *The Wealth of Nations*, the 1776 book that has become the Bible of free-market economics, Adam Smith wrote:

> Every individual endeavors to employ his capital so that its produce may be of greatest value. He generally neither intends to promote the public interest, nor knows how much he is promoting it. He intends only his own security, only his own gain. And he is in this led by an invisible hand to promote an end which was no part of his intention. By pursuing his own interest he frequently promotes that of society more effectually than when he really intends to promote it.[4]

This widely quoted statement is used as a justification for allowing business to manage its own affairs with a minimum of government interference.

But today, governments at state, federal and local levels are very much involved in regulating the environmental impact of American agriculture and industry. Environmental regulations are often complex and sometimes require specialists to interpret them. The costs of interpreting environmental laws and cleaning up the environment in accordance with the law add to the cost of the final product of farm or factory.

The government is now seeking ways to provide private industry with incentives to regulate its own environmentally hazardous behavior. Can the environment be saved by an "invisible hand"? Probably not. Nevertheless, even if government regulation cannot be eliminated altogether, perhaps it can be reduced. A free-market approach to environmentalism would reward businesses that reduce pollution and penalize those that add to pollution.

Project 88, a study sponsored by senators Timothy Wirth (D-Colo.) and John Heinz (R-Pa.) recommends applying market incentives in this way: The EPA would set emissions standards calling for reductions of, say, sulfur dioxide in order to combat acid rain. Factories that meet these standards would be awarded credits for their efficiency. A corporation, having some efficient factories and some older, more polluting factories can choose one of several responses. It can spend money on the old factories to bring them up to standard (building higher smokestacks, adding scrubbers, using more expensive fuel) and so earn as many as possible credits. Or it can apply the credits earned from its good factories to offset the demerits accumulated by the old factories.

An efficient factory can earn money by selling its credits to a less efficient factory. If it is in a position to reduce its emissions substantially and has many credits it does not immediately need, it can add substantially to its profits by selling credits. In this example, the profit motive—the pursuit of self-interest, or the invisible hand—works to help clean up the environment. In other industries, this plan could be modified. The principle, however, would remain unchanged. Agriculture and industry would find it in their self-interest to improve the environment.[5]

Other examples of how market forces might be made to work are:

- Western water shortages can be relieved by the sale of ownership rights to water. That is, farms or factories who have water to spare can sell to those who have too little.
- Wildlife poaching can be reduced by allowing ranches to establish private hunting preserves on their land. They would then have the incentive to monitor illegal hunting there.

- People can be given the incentive to generate less trash by charging them for the amount and type of trash they throw out.

Most environmentalists are unsure that these proposals can work. Nevertheless, the World Wildlife Fund (whose president, William Reilly, was appointed by President George Bush to head the EPA) has itself suggested a global market-oriented approach to one environmental problem. The plan is to relieve South American nations of a portion of their international debt burden if they protect rather than destroy their vast tropical rainforests. This approach is not only market-oriented, but it takes into account the need for global cooperation if the planet's environment is to be saved.

The Endangered Globe

If you were a space traveler, would you see the Earth as it really is? You would see that it is a planet of rare beauty, blue seas, sandy beaches, mighty mountains, and vast forests and grasslands. You would see it, as Preston Cloud wrote of it in *Cosmos, Earth, and Man*, as a "...relatively tiny habitable sphere, suspended in the black emptiness of the heavens, hurtling at 30 kilometers (about 18 miles) a second in captive and delicately balanced orbit around the life-giving, but potentially lethal, sun."[6]

In the well-known children's story, *Goldilocks and the Three Bears*, Goldilocks entered the bears' home and sampled their porridge. She found one portion too hot, one too cold, and a third "just right."

Earth is unique among the planets in being a fertile environment for life. Some planets are too far from the sun, and too cold for living things as we know them. Some are too close to the sun, and so hot nothing can live on them. Other planets have poisonous atmospheres, and still others lack water.

Of all the planets, Earth is "just right" for living things. But to keep it that way, a precarious balance must be maintained between Earth's water resources (the hydrosphere), its soil, rocks and mineral resources (the lithosphere), and its air (the atmosphere). Together, the three spheres are called the biosphere. We may try to expand the biosphere by sending rockets into space or by mining or drilling into the Earth's crust for coal, oil or natural gas, but the biosphere remains the relatively thin layer between earth and sky. It is the environment that makes plant, animal and human life possible.

Planet Earth is a "closed system." That is, the air and water now on Earth have always been there. Earth gets no new supplies from space. Whether it be a flea on your dog or a shark in the sea, all life in Earth's biosphere is an essential part of a delicate web. Every time you weed your garden to help roses grow, you change the biosphere in which you live. On a larger scale, the biosphere is changed when a condominium or golf course is built where a forest once stood. When people build a dam with which to generate electricity, construct a dike to hold back the sea, terrace a hill, or irrigate a desert, the web of life on Earth has been altered, for better or worse.

Until very recently, the principal forces of change in the biosphere were natural disasters such as volcanoes, forest fires, earthquakes, hurricanes, tornadoes, typhoons or tidal waves. In our own time, we are subject to technological disasters that can forever destroy the balance of the biosphere. These threaten plant, animal and human life.

In 1945, the United States dropped atomic bombs on the Japanese cities of Hiroshima and Nagasaki in order to hasten the end of World War II. In 1979, human error and equipment failure led to a nuclear accident at Three Mile Island in Pennsylvania. In 1984, deadly gas escaped from a chemical plant in Bhopal, India, and killed 2,500 people in the worst industrial accident in history. In 1986, at Chernobyl in the U.S.S.R.., a nuclear accident led to radioactive fallout over many cities of Europe.

Any disaster, whether made by people or by nature, changes the biosphere. The biosphere can usually bounce back from disaster fairly rapidly. It is delicate, but not easily destroyed.

But as man's power to abuse the system grows, it becomes more and more important to protect irreplaceable resources and find workable solutions to environmental problems.

The Environment Requires a Global Approach

There are some who might be called environmental activists. Greenpeace is an international organization established for the purpose of dramatizing danger to the environment. Two Greenpeace activists once risked their lives to scale a cliff at the junction of the Saguenay and St. Lawrence rivers in Quebec, Canada, in order to unfurl a huge banner emblazoned with the message "Sauvons Les Belugas"— "Save the Belugas" (a beluga is a species of whale). Thirteen Greenpeace members dropped over the railing of the Triborough Bridge in New York City and lowered themselves on ropes to hang 100 feet above Hell Gate, a narrow channel of the East River. After hanging a banner that protested ocean sludge dumping, they dangled in wait for eight hours. Forming a thin human curtain, they prepared to drop down to the water in the path of sludge barges attempting to pass under the bridge on the way to the sea.[7]

Whether or not these actions have any impact, they do dramatize the need for global action on environmental issues. According to the World Watch Institute, many global disasters are just waiting to happen because of shrinking tropical forests, soil erosion, falling underground water tables, rising global temperatures and sea level, and extinction of plant and animal species.[8]

National boundaries must not limit efforts to forestall these potential environmental disasters. For example, if the United States and other nations are going to halt the greenhouse effect, the destruction of forests in Brazil and Southeast Asia must be stopped. National self-interest must give way to the collective need for human survival. How can this be achieved?

In 1972, the United Nations took the first steps toward a global approach to environmental matters. It organized the Stockholm Conference, which in turn established the United Nations Environment Program (UNEP). In 1987, under the sponsorship of UNEP, 24 countries signed the Montreal Protocol calling for a reduction in the output of ozone-destroying chlorofluorocarbons. In Geneva, in 1988, 35 countries—including the United States—initiated an international effort to evaluate global climate trends. In particular, they sought to develop new approaches to the problem of global warming.

The rich, industrialized countries of the world are the worst polluters. Poor countries, meanwhile, are battered by problems of overpopulation, malnutrition, hunger and disease. Many of these poor nations have tried to industrialize rapidly, in order to improve the lives of their people. In doing so, they have frequently used their resources recklessly. They point out, quite correctly, that the United States developed its own industrial strength by denuding its forests and rapidly mining its coal. Is it not hypocritical, therefore, for the United States and other rich nations to criticize the approach poor nations are now taking to develop their own resources in search of a better life?

These nations have logic on their side, but not time. What was possible, if undesirable, 200 years ago is no longer possible today. By depleting their resources, underdeveloped nations not only mortgage their own futures, they endanger the future of all other people on planet Earth.

To accommodate the need of these nations for a better life while protecting the environment, global financial institutions such as the World Bank and the International Monetary Fund need to monitor the environmental impact of the funds they make available for development. "What is needed," said Senator Albert Gore of Tennessee, "is a Marshall Plan for economic development and environmental programs."[9] (The Marshall Plan, named after U.S. Secretary of State George Marshall, was a program of aid to Western Europe designed to help it recover from World War II.) Whether the nations of the world can join forces to finance such a program cooperatively remains to be seen.

Notes

Citations in the notes are brief. Full citations appear in the bibliography.

CHAPTER EIGHT

1. U.S. Department of Interior, 5.
2. U.S. Department of Interior, 43.
3. Naj.
4. Smith.
5. *Economist*, "The Greening of the Invisible Hand," December 24, 1988, 107–108.
6. Cloud, 1.
7. Harwood, 71–74.
8. Brown, 6.
9. Sancton, 63.

Epilogue: *An Environmental Ethic*

Every eight seconds a new American is born: He will need in his lifetime 56 million gallons of water, 21,000 gallons of gasoline, 10,000 pounds of meat, 28,000 pounds of milk and cream, and 9,000 pounds of wheat. In those terms, the American baby is 50 times more of a disaster to the environment of earth than is an East Indian....Every American consumes 3 tons of coal, 2 1/2 tons of cement, one ton of wood, 1/2 ton of nonferrous metals, 1/3 ton of steel, and over 1,500 liters of oil a year....Americans are the greatest resource-gobblers in the world....

So commented Stanley Beaver in J. Wreford Watson's book, *The United States: Habitation of Hope.*[1] What is wrong is living beyond one's means. To the extent that Americans and others in the rich countries of the world are living beyond their environmental means, they endanger the global environment. To the extent that they borrow resources to live well today at the expense of future generations, their reckless exploitation of resources needs to be checked. But how to do so?

Local, national and international effort must be brought to bear so that the world's resources are more equitably shared, and so that the future of the Earth is safeguarded. But these remedies will not be enough. People must learn a new environmental ethic and internalize a sense of their relationship to the environment.

Aldo Leopold first recognized that it would never be enough to preserve the environment through the force of law alone. Instead, something must happen in the human heart. As he wrote in *A Sand County Almanac and Sketches Here and There,* "people must examine each [environmental] question in terms of what is ethically and esthetically right, as well as what is economically expedient. A thing is right when it tends to preserve the integrity, stability, and beauty of the biotic community. It is wrong when it tends otherwise."[2]

Notes

Citations in the notes are brief. Full citations appear in the bibliography.

EPILOGUE

1. Watson, 326–27.
2. Leopold, 226.

Bibliography

Adams, James Truslow, *The Epic of America*, Boston: Little Brown and Company, 1931.

Andrus, Cecil. "In Contempt of History." *Wilderness* 46 (Spring 1983).

*Barnet, Richard J. *The Lean Years: Politics in the Age of Scarcity*. New York: Simon and Schuster, 1980.

Boorstin, Daniel. *The Americans: The National Experience*. New York: Random House, 1965.

*Brown, Lester R., and Edward C. Wolf. *Soil Erosion: Quiet Crisis in the World Economy* (World Watch Paper 60). Washington, D.C.: World Watch Institute, 1984.

Brown, Lester R., et al. *State of the World: A World Watch Institute Report on Progress Toward a Sustainable Society*. New York: W. W. Norton, 1988.

Canby, Thomas. "Our Most Precious Resource: Water." *National Geographic* 158 (August 1980).

*Carson, Rachel. *Silent Spring*. Boston: Houghton Mifflin Company, 1962.

Cloud, Preston. *Cosmos, Earth and Man*. New Haven: Yale University Press, 1978.

Cobb, Charles E., Jr. "The Great Lakes Troubled Waters." *National Geographic* (July 1987).

*Commoner, Barry. *The Closing Circle*. New York: Alfred A. Knopf, 1971.

Cruver, Philip C. "Lighting the 21st Century." *Futurist XXIII* (January/February 1989).

D'Allesandro, Bill. "Dark Days for Solar." *Sierra* 72 (July/August 1987).

Environmental Protection Agency. *Waste Minimization*. Washington, D.C., 1987.

Garland, Hamlin. *A Son of the Middle Border*. 1917. Reprint. Chattanooga, Tenn.: American Biography Service, 1988.

George, Henry. *Progress and Poverty*. Reprint. Sugar Hill, N.H.: Phoenix Publishing, 1979.

Gordon, Mitchell. *Sick Cities*. New York: Macmillan Company, 1963.

Grove, Noel. "Air: An Atmosphere of Uncertainty." *National Geographic* 171 (April 1987).

Halberstam, David. *The Reckoning*. New York: William Morrow and Company, 1986.

Hanson, Dennis. "Earthlog." *Audubon* 80 (November 1978).

Harwood, Michael. "Daredevils for the Environment." *New York Times Magazine*, October 2, 1988.

Hayes, Dennis. *Pollution: The Neglected Dimensions* (World Watch Paper #27). Washington, D.C.: World Watch Institute, 1979.

Hedley, Don. *World Energy: The Facts and the Future*. New York: Facts On File, 1986.

Herber, Lewis. *Crisis in Our Cities*. Englewood Cliffs, N.J.: Prentice Hall, 1965.

Holmes, Ralston, II. *Environmental Ethics*. Philadelphia: Temple University Press, 1988.

Johnson, Tom. "Lessons of Chernobyl." *New Republic* 14 (July 14, 1986).

*Leinwand, Gerald. *Air and Water Pollution*. New York: Pocket Books, 1969.

*Leopold, Aldo. *A Sand County Almanac and Sketches Here and There*. New York: Oxford University Press, 1949.

Lerner, Max. *America as a Civilization*. New York, Simon and Schuster, 1957.

Levin, Michael H. "The Trash Mess Won't Be Easily Disposed Of." *Wall Street Journal*, (December 15, 1988).

MacFayden, J. Tevere. "Where Will All the Garbage Go?"*Atlantic* (March 1985).

Marx, Wesley. "Swamped by Our Own Sewage." *Reader's Digest* (January 1988).

Matthiessen, Peter. "Our National Parks: The Case for Burning." *New York Times Magazine* (December 11, 1988).

Naj, Amal Kumar. "How to Clean Up Superfund's Act." *Wall Street Journal* (September 15, 1988).

Postel, Sandra. *Water: Rethinking Management in an Age of Scarcity.* Washington, D.C.: World Watch Institute, 1984.

Powledge, Fred, "Water, Water, Running Out." *Nation* 234 (June 12, 1982).

Sancton, Thomas A. "Hands Across the Sea." *Time* (January 2, 1989).

Schlesinger, Arthur M. Jr. *The Age of Roosevelt: The Coming of the New Deal.* Boston: Houghton Mifflin Company, 1959.

Sheaffer, John R., and Leonard A. Stevens. *Future Water.* New York: William Morrow and Company, 1983.

Sherman, David. "The Underwatered West: Overdrawn at the Well." *Environment* 23 (March 1981).

Smith, Adam. *The Wealth of Nations.* New York: Penguin Books, 1982.

Steinhart, Peter. "Down in the Dumps." *Audubon* 88 (May 1986).

Udall, Stewart. "The Land Wisdom of the Indians." In *Environmental Decay in Its Historical Context,* by Robert Detweiler et al. Glenview, Ill.: Scott, Foresman and Company, 1973.

U.S. Department of Agriculture. *Conquest of the Land through Seven Thousand Years.* Agriculture Information Bulletin no. 99. Prepared by W. C. Lowdermilk. Washington, D.C., revised 1986.

U.S. Department of the Interior. *The Department of Everything Else: Highlights of Interior History.* Prepared by Robert M. Utley and Barry Mackintosh. Washington, D.C., 1988.

Watson, J. Wreford. *The United States: Habitation of Hope.* London: Longman, 1982.

Watkins, T. H. "For a Measure of Wisdom." *Wilderness 46* (Spring 1983).

Weber, Bruce. "Overtired," *New York Times Magazine* (February 7, 1988).

Weinberg, David. "An Unhealthy Mixture of Science and Politics." *Natural History Magazine* 92 (October 1983).

White, Lynne, Jr. "The Historical Roots of our Ecological Crisis." In *Environmental Decay in its Historical Context,* by Robert Detweiler et al. Glenview, Ill.: Scott, Foresman and Company, 1973.

*Especially recommended for further reading

INDEX